Georgios Miaoulis and Dimitri Plemenos (Eds.)

Intelligent Scene Modelling Information Systems

Studies in Computational Intelligence, Volume 181

Further volumes of this series can be found on our homepage: springer.com

Vol. 158. Marina Gavrilova (Ed.)
Generalized Voronoi Diagram: A Geometry-Based Approach to Computational Intelligence, 2009
ISBN 978-3-540-85125-7

Vol. 159. Dimitri Plemenos and Georgios Miaoulis (Eds.)
Artificial Intelligence Techniques for Computer Graphics, 2009
ISBN 978-3-540-85127-1

Vol. 160. P. Rajasekaran and Vasantha Kalyani David
Pattern Recognition using Neural and Functional Networks, 2009
ISBN 978-3-540-85129-5

Vol. 161. Francisco Baptista Pereira and Jorge Tavares (Eds.)
Bio-inspired Algorithms for the Vehicle Routing Problem, 2009
ISBN 978-3-540-85151-6

Vol. 162. Costin Badica, Giuseppe Mangioni, Vincenza Carchiolo and Dumitru Dan Burdescu (Eds.)
Intelligent Distributed Computing, Systems and Applications, 2008
ISBN 978-3-540-85256-8

Vol. 163. Pawel Delimata, Mikhail Ju. Moshkov, Andrzej Skowron and Zbigniew Suraj
Inhibitory Rules in Data Analysis, 2009
ISBN 978-3-540-85637-5

Vol. 164. Nadia Nedjah, Luiza de Macedo Mourelle, Janusz Kacprzyk, Felipe M.G. França and Alberto Ferreira de Souza (Eds.)
Intelligent Text Categorization and Clustering, 2009
ISBN 978-3-540-85643-6

Vol. 165. Djamel A. Zighed, Shusaku Tsumoto, Zbigniew W. Ras and Hakim Hacid (Eds.)
Mining Complex Data, 2009
ISBN 978-3-540-88066-0

Vol. 166. Constantinos Koutsojannis and Spiros Sirmakessis (Eds.)
Tools and Applications with Artificial Intelligence, 2009
ISBN 978-3-540-88068-4

Vol. 167. Ngoc Thanh Nguyen and Lakhmi C. Jain (Eds.)
Intelligent Agents in the Evolution of Web and Applications, 2009
ISBN 978-3-540-88070-7

Vol. 168. Andreas Tolk and Lakhmi C. Jain (Eds.)
Complex Systems in Knowledge-based Environments: Theory, Models and Applications, 2009
ISBN 978-3-540-88074-5

Vol. 169. Nadia Nedjah, Luiza de Macedo Mourelle and Janusz Kacprzyk (Eds.)
Innovative Applications in Data Mining, 2009
ISBN 978-3-540-88044-8

Vol. 170. Lakhmi C. Jain and Ngoc Thanh Nguyen (Eds.)
Knowledge Processing and Decision Making in Agent-Based Systems, 2009
ISBN 978-3-540-88048-6

Vol. 171. Chi-Keong Goh, Yew-Soon Ong and Kay Chen Tan (Eds.)
Multi-Objective Memetic Algorithms, 2009
ISBN 978-3-540-88050-9

Vol. 172. I-Hsien Ting and Hui-Ju Wu (Eds.)
Web Mining Applications in E-Commerce and E-Services, 2009
ISBN 978-3-540-88080-6

Vol. 173. Tobias Grosche
Computational Intelligence in Integrated Airline Scheduling, 2009
ISBN 978-3-540-89886-3

Vol. 174. Ajith Abraham, Rafael Falcón and Rafael Bello (Eds.)
Rough Set Theory: A True Landmark in Data Analysis, 2009
ISBN 978-3-540-89886-3

Vol. 175. Godfrey C. Onwubolu and Donald Davendra (Eds.)
Differential Evolution: A Handbook for Global Permutation-Based Combinatorial Optimization, 2009
ISBN 978-3-540-92150-9

Vol. 176. Beniamino Murgante, Giuseppe Borruso and Alessandra Lapucci (Eds.)
Geocomputation and Urban Planning, 2009
ISBN 978-3-540-89929-7

Vol. 177. Dikai Liu, Lingfeng Wang and Kay Chen Tan (Eds.)
Design and Control of Intelligent Robotic Systems, 2009
ISBN 978-3-540-89932-7

Vol. 178. Swagatam Das, Ajith Abraham, Amit Konar
Metaheuristic Clustering, 2009
ISBN 978-3-540-92172-1

Vol. 179. Mircea Gh. Negoita, Sorin Hintea
Bio-Inspired Technologies for the Hardware of Adaptive Systems, 2009
ISBN 978-3-540-76994-1

Vol. 180. Wojciech Mitkowski, Janusz Kacprzyk (Eds.)
Modelling Dynamics in Processes and Systems, 2009
ISBN 978-3-540-92202-5

Vol. 181. Georgios Miaoulis, Dimitri Plemenos (Eds.)
Intelligent Scene Modelling Information Systems, 2009
ISBN 978-3-540-92901-7

Georgios Miaoulis
Dimitri Plemenos
(Eds.)

Intelligent Scene Modelling Information Systems

Georgios Miaoulis
Technological Education Institution of Athens
Dept. Computer Science
Ag. Spyridonos Str.
122 10 Athens
Egaleo
Greece

Dimitri Plemenos
Universite Limoges
Laboratoire MSI
83 rue d'Isle
87000 Limoges
France
Email: plemenos@unilim.fr

ISBN 978-3-540-92901-7 e-ISBN 978-3-540-92902-4

DOI 10.1007/978-3-540-92902-4

Studies in Computational Intelligence ISSN 1860949X

Library of Congress Control Number: 2008942135

Typeset & Cover Design: Scientific Publishing Services Pvt. Ltd., Chennai, India.

Printed in acid-free paper

9 8 7 6 5 4 3 2 1

springer.com

Preface

Scene modeling is a very important part in Computer Graphics because it allows creating more or less complex models to be rendered, coming from the real world or from the designer's imagination. However, scene modeling is a very difficult task, as there is a need of more and more complex scenes and traditional geometric modelers are not well adapted to computer aided design. Even if traditional scene modelers offer very interesting tools to facilitate the designer's work, they suffer from a very important drawback, the lack of flexibility, which does not authorize the designer to use incomplete or imprecise descriptions, in order to express his (her) mental image of the scene to be designed. Thus, with most of the current geometric modelers the user must have a quite precise idea of the scene to design before using the modeler to achieve the modeling task. This kind of design is not really a computer aided one, because the main creative ideas have been elaborated without any help of the modeler.

Declarative scene modeling could be an interesting alternative to traditional geometric modeling. Indeed, declarative scene modeling tries to give intuitive solutions to the scene modeling problem by using Artificial Intelligence techniques which allow the user to describe high level properties of a scene and the modeler to give all the solutions corresponding to imprecise properties.

No matter if a scene modeler is declarative or a classical geometric one, it is usually not enough to help the user in the designing process. The designer has not always to start from nothing and to design a new scene. Scene data bases, or scene knowledge bases in the case of declarative modeling, should be associated to the scene modeler, in order to allow the designer to use and/or to modify existing models during the designing process. Moreover, easy access to Internet, together with appropriate tools, should be available for the user, allowing, for example, collaborative scene modeling. In other words, scene modeling tools should be complete information systems, that is *scene modeling information systems*.

This book is dedicated to *intelligent scene modeling information systems*, that is information systems using Artificial Intelligence techniques to design scenes. Declarative scene modeling techniques are presented, as well as their implementation in an intelligent information system.

In order to improve efficiency of declarative modeling based scene modeling information systems, various techniques are proposed in the book: coupling of a declarative modeler with a classical geometric modeler. Use of machine-learning Artificial Intelligent techniques, allowing the system to learn the user preferences; introduction of high level concepts, especially the concept of "style" in architectural design; introduction of collaborative declarative modeling techniques.

The last chapter is dedicated to the web security problem, as an intelligent scene modeling information system must be able to work in open environments. Artificial Intelligence techniques, as well as information visualization techniques, are proposed to easily detect web attacks.

Intelligent scene modeling information systems may reduce the complexity of scene modeling, by using well adapted Artificial Intelligence techniques, together with a lot of libraries and other tools, allowing easy description, search or modification of scenes. The use of declarative modeling techniques in the heart of such systems was experimented and seems to give interesting results, even if these techniques have to be still improved.

Georgios Miaoulis
Dimitri Plemenos

Contents

1
Intelligent Scene Modelling Information Systems: The Case of Declarative Design Support

Georgios Miaoulis

Department of Informatics, Technological Education Institute of Athens,
Ag.Spyridonos St., 122 10 Egaleo, Greece
gmiaoul@teiath.gr

Abstract. The current chapter presents and analyses an intelligent information system dedicated to scene modelling using the declarative design by hierarchical decomposition methodology. The aim is to develop a set of rules and standards, which constitute a functional process and data framework covering a wide range of similar systems. The basic mechanisms of the specific design process models are exposed. The scene representation methods are defined combined with a strong semantic dimension of the system and a three abstraction layers norm. A special scene database system within a dedicated informational and knowledge environment are exposed. The software system architecture of such a range of systems is defined. Finally, a pilot software system is used in order to concretise its results.

Keywords: Intelligent Information Systems, Scene Modelling, Declarative Design, Design Support Systems, Scene Data Bases.

1.1 Introduction

The "Intelligent Scene Modelling Information System" or ISMIS, as an acronym, is a neologism that determines a special *information system* which deals with (a) the use of *artificial intelligence techniques* and *knowledge management* (b) the use of different kinds of *visual information* such as pictures, image synthesis and visual information in general (c) the application of *interactive communication* techniques and (d) *collaborative* possibilities. If we specify in this description the term "*Scene Modelling*" as a specific process orientation (modelling) and data orientation (scene in image synthesis) that characterise the aforementioned system, we obtain the difference between an *Intelligent Visual Information System* and an ISMIS.

For the study of the abovementioned systems many scientific fields such as *image processing*, *computer graphics, data visualisation, computer vision*, *digital media* as well as *visual communication, human computer interaction*, *computational intelligence* and *knowledge-based systems* contribute and crossover in the context of the applied discipline.

The application fields of such systems are various. We can find many applications that obey this profile: for example *Computer-aided diagnosis* based on intelligent image understanding, *Computer assisted learning* based on verbal and visual information and use artificial intelligence techniques or simulations.

G. Miaoulis and D. Plemenos (Eds.): Intel. Scene Mod. Information Systems, SCI 181, pp. 1–27.
springerlink.com

Creative support systems represent a prominent application field of developing this kind of systems. This family of systems can support the user by encouraging exploration, enhancing collaboration, transferring heavy combinatorial computation tasks to computer systems in order to produce and investigate alternative solutions.

The *Design Support Systems* ranging from simple geometric modellers for drawing applications to the most sophisticated creative support environments, which are influenced from advanced design methodologies, demonstrate the evolution and the openness of this particular domain.

In order to clarify the relevant issues, we explore the main dimensions which characterise these systems, e.g. *Intelligence*, *Visual Information and Perception*, *Interactivity* and *Collaborativity* in the frame of an information system.

Integrated *Intelligent Information Systems* typically comprise the implementation of four equally important components or stages [3, 28, 29],

- Real world information is transformed into data, stored and processed by a computational environment. This stage may be connected with sensor technology, analogue-to-digital conversion, sampling techniques, as well as input explicitly formulated by humans.
- Already existing knowledge regarding the concept to be learnt is codified as a set of rules by human experts.
- New experiences or facts, appropriately transformed to data, are exploited in order to advance the intelligent system's behaviour. This stage is connected with machine learning methodologies and algorithms.
- The system takes advantage of the outcome of the aforementioned tasks in order to exhibit intelligent behaviour.

These tasks are typically supported by an infrastructure for data and knowledge storage management and communication with the user.

The field of *visual information and perception* [12] characterises an increasing range of modern information systems. This type of systems uses interactive visualisation with special processing techniques to respond to a wide range of decision problems in art, science, engineering or management. The visual dimension of these systems uses techniques from computer graphics, data visualization, perception and cognition. A visual information system has as principal aspect the visual representation of output data. Data may contain pictorial or graphical information or raw data in textual form and the transformation of this data in visual representations is the key future of these systems.

The degree of *visual interactivity* is also another dimension appertaining to such a system. Alternatively, visual interactivity may be expressed by the ratio of initiative of the user to the visual stimuli offered by the system. This initiative or reaction is based on the *visual analytics* capabilities developed through the coupling of specialised visual representation, analysis computer tools and the human analytical reasoning skills.

The degree of *collaborativity* expresses the level of capability of the aforementioned design environments to support both individuals' and groups' creativity. Collaborative design activities are the main challenge for this type of systems. Shared spaces, wikis, resources and media sharing constitute examples of implementations which may be incorporated into dedicated systems or may act independently.

The following example evaluates five "typical" systems in the four aforementioned dimensions in order to illustrate the nature of an Intelligent Visual (or Scene Modelling) Information System.

Table 1.1. Evaluation of five "typical" information systems

	Intelligence	Visual perception	Interactivity	Collaborativity
IVIS	High	High	High	High
New Media	Low	High	High	Low
Intelligent	High	Low	Midium	Low
Visual	Low	High	Low	Low
Conventional	Low	Low	Medium	Low

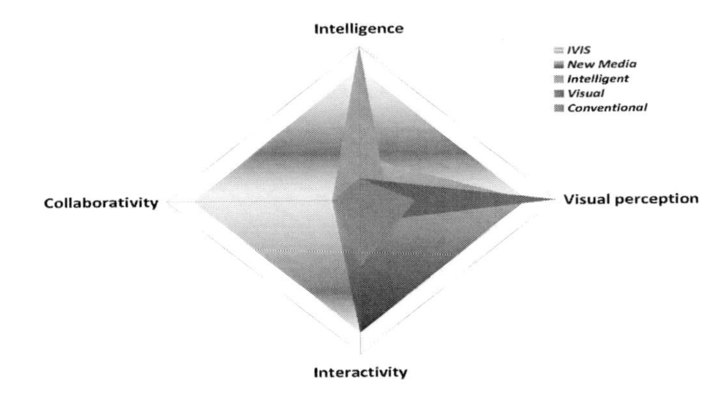

Fig. 1.1. The area of Intelligent Visual Information Systems

In this chapter we try to analyse the major characteristics of an ISMIS through a particular case study. This case study concerns a *Declarative Design Support System* (DDSS) or a *Computer Assisted Declarative Design* (CADD) system [23, 24] comprising an appropriate environment for design by according to the *Declarative Scene Modelling* Methodology (DSM) [8, 20, 21]. For this system, we have also developed a pilot implementation called **MultiCAD** [24, 25] employing the *Declarative Modelling by Hierarchical Decomposition* (DMHD) [27] version of this methodology.

1.2 The Scene Modelling Process in Declarative Design Support

The nature of the *design process* [14] corresponds to a collection of multilateral mental activities, [1, 2], highly complex, taking place in a *non-linear* and *iterative* manner. The prototype forms need to be constructed, evaluated and, subsequently, reformulated in order to offer the necessary understanding for the next, more elevated and more refined level of the solution. This notion of *incrementalism* (by subsequent refinements) [31] is based on the assumption of limited information processing capacity on behalf of the decision maker (or the designer in the specific context).

The unfolding of the design process may present a large variety of procedural structures (or mechanisms) and, therefore, may not be limited by a particular methodology. The design process in its entirety bears all different *phases*[1] and *stage* that form the *process lifecycle* [7].

Analysis of the Declarative Modelling Design Paradigm

The core of declarative modelling connect to the idea that "we can understand the world in other ways, apart from its geometric description: we can perceive it by its properties, by its characteristics, i.e. not only by the appearance presented to us but also by the mechanisms and the constraints upon which this form has been based." [20].

The process apparently requires a solution exploration engine. This engine must be able to interpret the declarative description language and produce a set of solutions conforming to the requirements of the initial model.

The design of physical entities (occupying space) through declarative modelling suggests a design process with a more or less strict sequence of stages. The activities of these stages are the description, the solution search and the understanding. The relations among these basic phases appear in Figure 1.2.

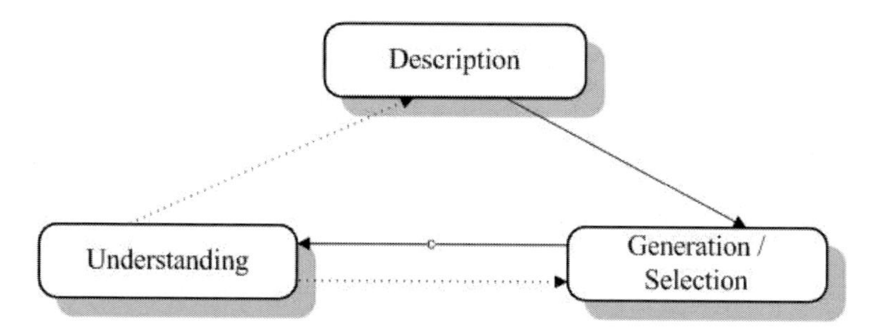

Fig. 1.2. Declarative Modelling basic phases cycle

Towards an Enhanced Declarative Modelling Paradigm

The comparison of various examples allows us to formulate remarks in regard to the redefinition of the activities of the declarative modelling paradigm and to propose corresponding process models, integrating, among others, knowledge, cases and information aspects [32]. Figure 1.3 presents the decomposition of the set of functions of the declarative design system.

Activity Modelling

Ideas, Needs, Rationale Activity: This activity corresponds to the clarification of the design problem, the recording of needs, and the formation of initial ideas leading to the construction of a description of high level of abstraction. In this activity there is the notion of planning of the continuity of the design, the project organization, the allocation of documentation or other resources for the accomplishment of its mission.

---

[1] Applicable mainly in Engineering.

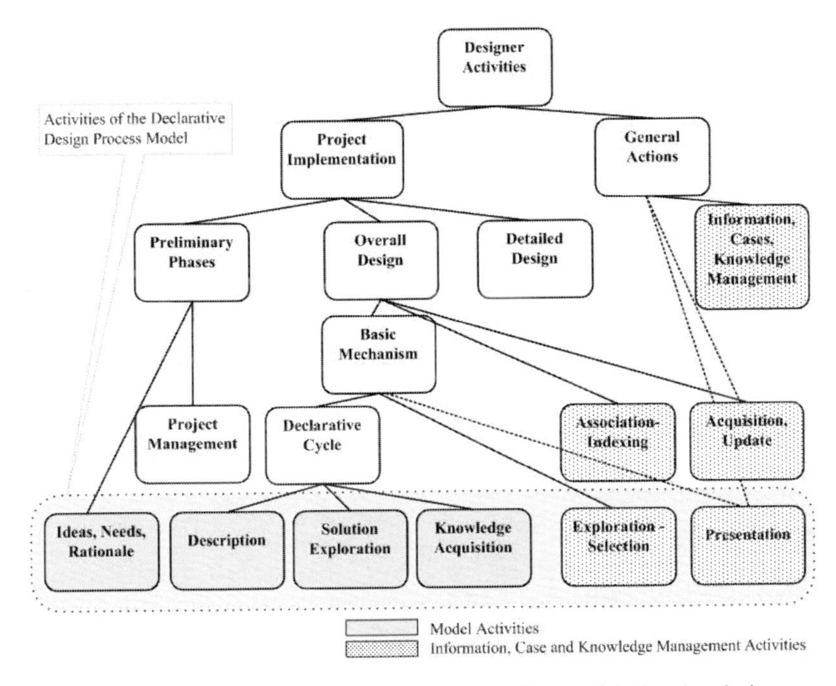

Fig. 1.3. Decomposition of functions related to the act of declarative design

Enhanced Description Activity: We have already underlined that the particularity of the description activity in declarative modelling is the construction of "complete" models of the product at a higher level of abstraction (and detail), which still remain however product models. Hence, declarative modelling possesses all the characteristics of an autonomous (or semi-autonomous) act of design. We must, therefore, include all necessary tasks at its level of abstraction in order to achieve a design fulfilling the declarative models. During this activity, we may also explore previous cases and corresponding information and proceed to the adaptation of the solutions for our current design problem. The description may also be distinguished in *initial description* and *adaptation of the (initial) description.* The content of these two types of description may vary significantly.

This aspect (autonomous design) inevitably increases the complexity of the description process. If we accept as a basis the framework described below, we may propose the following sub-activities:

1. Analysis
2. Specifications ⇔ *Synthesis (Construction of the declarative model).*
3. Refinement[2] ⇔ *Identification of anomalies, critique, evaluation.*
4. Information Retrieval ⇔ *Retrieval of cases and information.*

Figure 1.4 presents a possible model for this activity, corresponding to the decomposition and the above analysis.

[2] Elimination of problems introduced to the declarative model during the synthesis tasks.

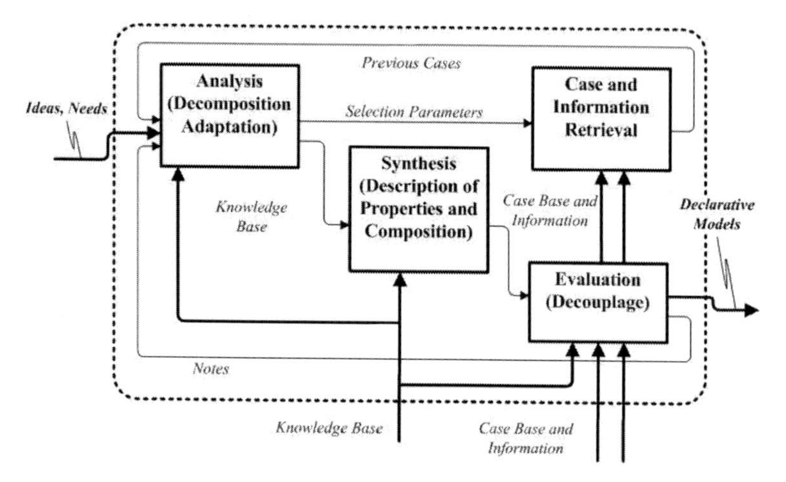

Fig. 1.4. The Enhanced Description Activity model

Generation/Selection Activity: The stage of *solution generation* comprises the possibility of the production of a solution space (using different types of *generation engines*). This production could be automatic or may require the designer's intervention (*interactive guided generation*). These solutions respect the constraints of the initial description (and those which were subsequently added during the generation). The **generation**, therefore, is equivalent to the *composition* and the definition of *property values* of the solution components (with respect to the declarative description specifications).

This stage could take place by **selection** from a *case base* or a *solution population* produced by using less restrictive specifications. This presupposes an information system dedicated to the declarative design and an *exploration engine* (of specific processing).

Knowledge Acquisition and Evaluation Activity: The operations covered by this activity could be multiple aiming to (a) *acquire knowledge* and insight from an individual solution or a group of solutions, (b) *select* among several solutions, (c) offer the ability to *refine* these solutions [12].

Exploration and Management of Knowledge, Cases and Information Activity: The activities of management of information resources, knowledge and cases[3] necessitates the presence of activities (or processing) which are either autonomous or incorporated in other functions of the design process.

Process lifecycle aspects

We have adopted the assumption that the design process (declarative or other) evolves according to two movements towards its final product. One movement is cyclic and we call it *basic cycle* or *design stage* which is decomposed in elementary activities. The other movement, connected with the displacement towards the final product, is linear and we call it *evolution of the design process.*

---

[3] Here, the term *case* covers both the notion of a *design project* as well as a *component,* a single *object.*

The declarative design process follows the general sequential scheme of phases that we have described above. It is focused on the phase of *embodiment design,* after the preliminary phases and before the phase of detailed design. Hence, the sequential aspect of the process adheres to this schema. The design process also follows a path leading from the general to the specific, from the abstract to the concrete. On the other side is the activity of the formation of *ideas, needs and the rationale* of the design. The stages of decomposition of the design process are similar and decomposable stages which may also comprise four elementary activities.

The mechanism model of the basis of the declarative paradigm constitutes the basic cycle (description – generation – knowledge acquisition) of the iteration.

The alternative cycles (generation – selection): The introduction of the notion of the solution exploration type, i.e. by generation and by selection of already generated or stored solutions leads us to suppose that the path of the basic cycle could follow different routes: (a) description – generation – knowledge acquisition, (b) description – selection – knowledge acquisition, (c) The *combined* type (description – generation – selection – knowledge acquisition) or even (description – generation – knowledge acquisition – selection) (d) The *alternating* type, i.e. one or more generation stages followed by selection stages.

Multilevel Iterations: The existence of iterative mini-cycles within the frame of each activity of the process is, practically, very important for the design of software for interactive CADD.

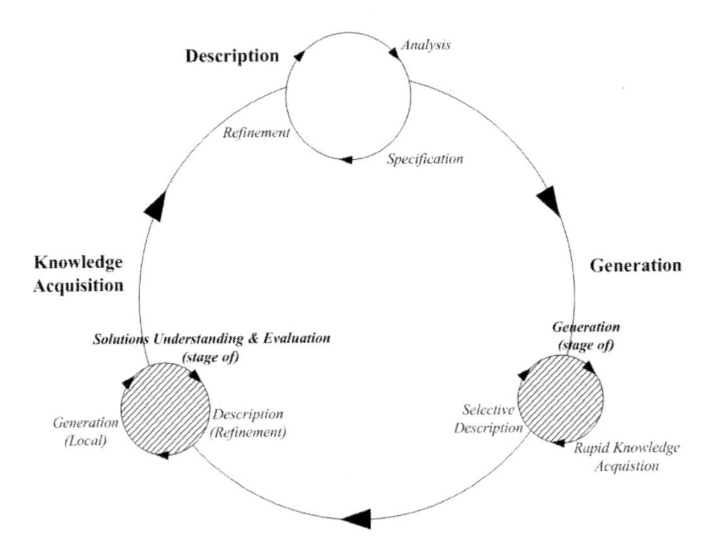

Fig. 1.5. The iterations in multiple levels

The exploitation of these mini-cycles contributes, essentially, the cooperative aspect of these software applications Figure 1.5 presents this aspect of multilevel iteration of the declarative process.

Process Evolution Curve: The design process is defined as a sequence of phases and stage-types of design, in regard to the simulation of its unfolding. This sequence of phases and stages could be represented by a spiral in a three-dimensional space defined by three axes [33] (Figure 1.6):

The **abstraction** axis.
The **detail level** axis.
The **lifecycle** axis.

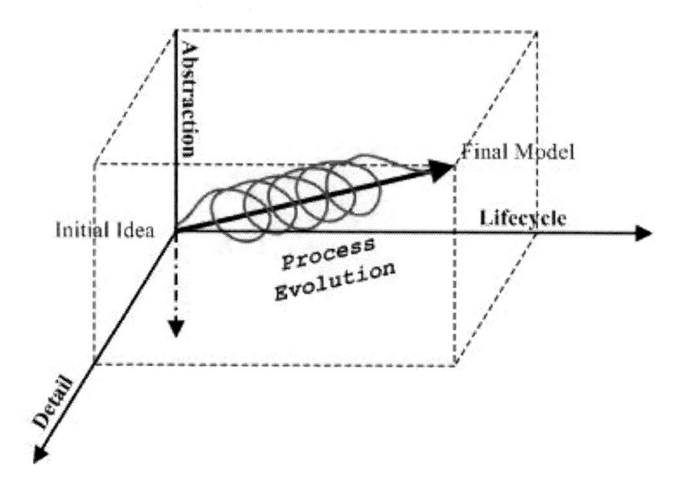

Fig. 1.6. The design process evolution

Generally, the level of detail increases and the level of abstraction decreases during time, until a complete model is obtained.

Declarative Design Process Model

In this section, we have employed a model of IDEF type [11] in order to describe the flow of information in the entire set of processes of declarative design. The diagram of this model is presented in Figure 1.7.

Functional Framework of Support Software System

In the previous sections, we have developed a series of models in order to cover the different aspects of the functionality of the declarative design support system. This *proposed functionality* relies on the principles of declarative modelling by hierarchical decomposition and the potential of software utilities and constitutes the *Support software functional framework* named also **"MultiCAD framework"**. Moreover it contains directive lines of the specifications for the development of *functional components* of the information system and his framework architecture.

The aforementioned aspects concern:

1. The different functions implied in such a system and the decomposition of these functions into elementary activities.
2. The content of these activities, their execution mode, their resources, their product and the flow of information and knowledge transformed by them.
3. The factors influencing the workflow of the designer, the modelling of the iterations.
4. The framework of usage of software utilities, the interaction modes.

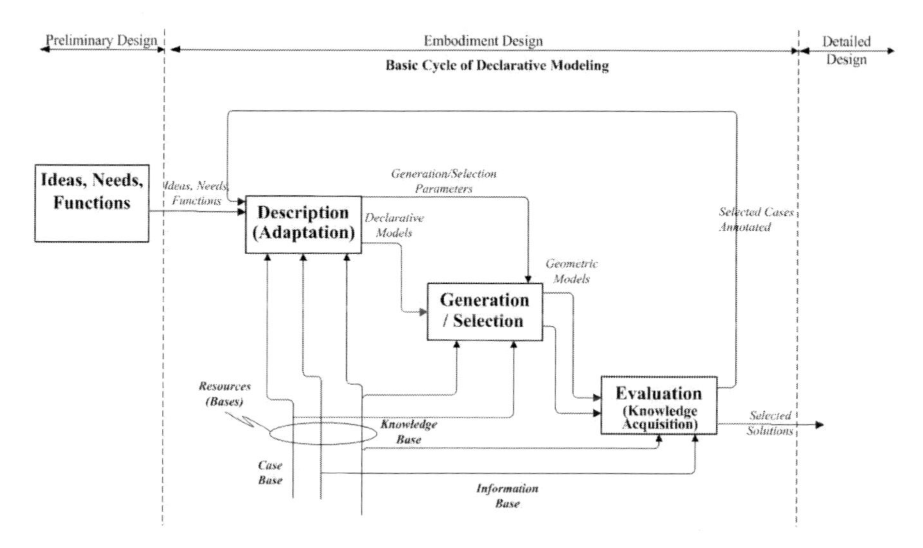

Fig. 1.7. Declarative Design Process Model

The proposed functional framework aims to contribute:

- To the evolution of the development of declarative models towards *a true autonomous design,* supported by appropriate software utilities and by a dedicated information system and knowledge.
- To the establishment of the necessary conditions for the development of *cooperative software* covering the design cycle in its entirety.
- To the development of the appropriate utilities for the activities related to *usage of resources* for knowledge, information and designers' logs.
- To the preparation of the necessary environment for the *integration of the design process* and the subsequent employment of *multiple design approaches.*

1.3 Information, Knowledge and Scene Models Representations

The information an ISMIS is required to handle may vary in nature as well as in volume, ranging from a few plain text and numerical tables to large numbers of high resolution still images, scene models or videos. The organisation and manipulation of the corresponding data is of utmost importance for the efficiency of such a system and, in cases of increased data volumes, even for the feasibility of its implementation. In the following we demonstrate the differences in the nature of information and knowledge and the diversity of data that may be yielded in different contexts bearing different data storage and management requirements. Next, we choose on one of the two major categories – visual vs. non-visual information – and present the main issues that have to be addressed for this category. We focus specially in the case of scene representations in different level of abstraction and we highlight the semantic dimensions of those models. Finally, we describe a unified conceptual framework in the context of an DDSS (case of MultiCAD) [24, 25, 26], able to handle the hole set of heterogeneous data and knowledge in a uniform way.

Scene Models Content and Classification

Informational Content of Scene Models: The dominant information in the Multi-CAD DDSS is that concerning the *conceptual description of the objects* comprising the scene. Through various transformations, these conceptual object descriptions may result in *geometric models, images* or *documents.*

The entire informational content of a scene model can be summarized as follows:

> **Scene Models** = Models of visual objects (form+appearance) + synthesis structure + spatial arrangement + environment properties

The scene models are therefore comprised by *descriptions* of their *elements, operations* on their elements, as well as the *synthesis structure* and the *spatial arrangement* (i.e. the *topological* structure) of their elements. The element descriptions cover the *object form* (i.e. the *geometry* of the entity), *properties* like colour, texture or others affecting the *appearance of objects,* as well as *application-specific properties* and additional information, associated with the scene components, such as material characteristics or functional features. A complete scene model also has to include *elements of the environment* such as lighting sources, backgrounds, properties relevant to the point of view.

Axes of Classification of Scene Models: The models employed for scene representation may be distinguished between concrete and generic models. A *concrete model* is a well defined scene representation which may be visualized without ambiguity whereas a *generic model* represents a whole set of concrete scenes. The unambiguous character of a model is essentially connected with its *visualisation mode.* For example, visualisation in "wire frame" form does not require an unambiguous model containing *appearance properties*, such as colour or texture, in order to be visualized. However, the same model may yield a set of alternative realistic visualisations if we assign a series of alternative values to its appearance properties. A *generic model* is also an *abstract model.* Hence, we may have a definition of the level of abstraction utilizing the notion of a generic model.

A model is considered to be of inferior abstraction level than another if it bears an inferior *degree of ambiguity* (and, consequently, a more restricted range of generated solutions) or/and it contains more *categories* of *attributes, properties* or *relations.* Hence, each type of properties or relations, or group of types, forms an **axis of abstraction**. For example, the properties relevant to the *form of the objects*, the *synthesis* and the *spatial arrangement* yield an **axis of geometric abstraction**.

The **levels of detail** ensue from the decomposition process. In particular, a model is considered to belong to a more detailed level than a given level of detail, if it contains several objects (components) related through "part-of" relations with the objects of the model of the given level of detail. The set of these detail levels forms an axis of detail and of evolution of models.

The scene models are subjects to successive transformations and refinements, taking into account the iterative nature of the design process (basic cycle, mini cycles). These models could be recorded on the **time axis** (the lifecycle axis) of the process evolution and those of the representations. There exists, therefore, a series of pieces of information regarding the **versions**, the access, the designers that operate on a representation as well as the applied transformations.

Scene Representation Typology: In the framework of an intelligent scene modelling information system (ISMIS) several types of scene models exist. These models could be classified in a hierarchy of representations from the most abstract to the most concrete, on an *abstraction level axis,* and from the most general to the most detailed, on a *detail level axis,* which reflect the evolution of the object definition throughout the design process. The definition of these axes has already been discussed in the previous section. For the abstraction, we have chosen three predefined levels as follows:

Abstract models (**conceptual**)**:** The abstract models are those scene models of declarative or descriptive type, the natural language documents or even the models based on semantic networks. These models are necessarily generic and they are able to offer a series of models of lower abstraction levels.

Intermediate models**:** These are, essentially, **geometric** models. These models are concrete, from the point of view of object forms, their synthesis and their spatial arrangement. The geometric models may yield several different models of lower abstraction level if, for example, we vary appearance properties such as object colour or texture. They may also produce images of a scene from different points of view.

Physical models**:** The models of the physical level contain information which is appropriate for the direct visualisation of 3D objects. They contain this information either in a raster form (concrete graphics) or in a vector form (abstract graphics), but in both cases containing all the necessary properties for the close representation of real situations.

The description type in each abstraction level of an ISMIS is different. Moreover, even in the same abstraction level, we may have different model types, depending on the nature of the language used (graphical or textual language) or on the form of the model. The relations among the different types of scene representation and the abstraction levels appear in Table 1.2:

Table 1.2. Scene models classification by abstraction level

Representation Typology	Abstract (generic)	Intermediate	Physical
Symbolic	Declarative models Formal or "natural" languages Semantic networks Symbolic graphics		
Abstract Graphics	Sketch (2D) parametric models, variational (3D)	Structured objects (scene graphs), vector models, surfaces, volumes, etc.	Vector models, surfaces, volumes, structured objects
Concrete Graphics / pictures			Images (point of view)

Models Transformation and Equivalence: In the definition of the abstraction level, where we have utilised the terms generic and specific in order to clarify the concept of abstraction, we have introduced the notion of vertical transformation, by model abstraction levels. In case a generic model undergoes a **generation** (transformation of), it will yield a series of concrete models (or more concrete, less ambiguous). The **understanding** of the image is the inverse process of the generation. The principal tasks during the design process are the editing, visualisation and transformation of the scene representation in order to obtain the solution. We have two kinds of transformations: (a) The *inter-level* transformations (generation and understanding of images) and (b) The *intra-level* transformations (conversion) (Figure 1.8).

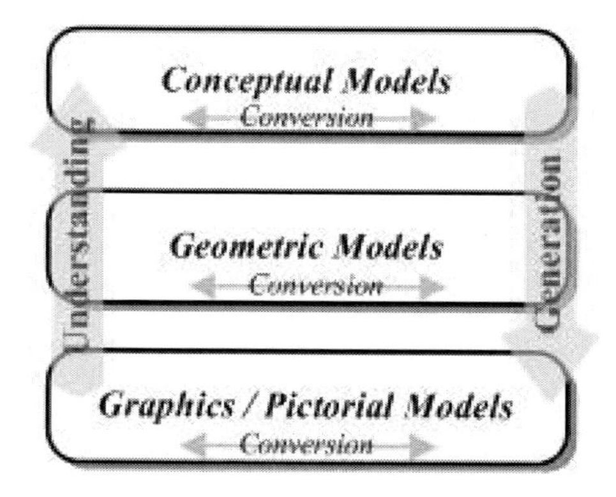

Fig. 1.8. The principal scene model transformations

Information associated with scenes: There exist various information elements relevant to the different aspects of the design product. These include, among others: the *functionalities* expected by the product, the *needs* covered by these functionalities and the connection with the conceptual description that could be established by the designer as well as another information structure, a hierarchy of importance for the design system. In a similar manner, the *explanation of the design decisions* is also important and constitutes crucial information with respect to the entire process and the corresponding product. The structure implied by these explanations could assist all tasks related to the modification or reutilisation of a scene. Another aspect concerns the *non-geometrical information* connected with a scene or a product. As an example, this information could refer to the physical properties of an object.

1.3.1 Physical Scene Models

The scene representations in physical level consist also of element descriptions and operations upon these elements. These elements concern the object form, the colour or other properties, as well as elements of the environment such as light sources or others. A graphical representation will be demonstrated through the production of an *image,* using the appropriate mathematical descriptions and operations which are known to the *visualisation engine.*

The scenes may be distinguished in *two-dimensional* and *three-dimensional* scenes. The different scene representations are based on models. These models are *geometric models* (composed by *simple elements*, *surfaces* or *volumes*), *models based on physics* (e.g. *fractal models*) and *algorithmic models* (e.g. LOGO).

The description of different (model based) scene representations and their common operations demonstrate the fact that the degree of structuring allows editing operations at a high level (similar to textual representations) while being independent of the rendering engine itself. The appearance of an entire scene may vary considerably from one rendering engine to another.

Graphs and Networks based representations

The *scene graphs* (figure 1.9.) are representations of ordered sets of object named nodes (3D forms, properties or groups), which represent the elements composing a scene. A scene graph may be a tree or a directed acyclic graph (DAG).

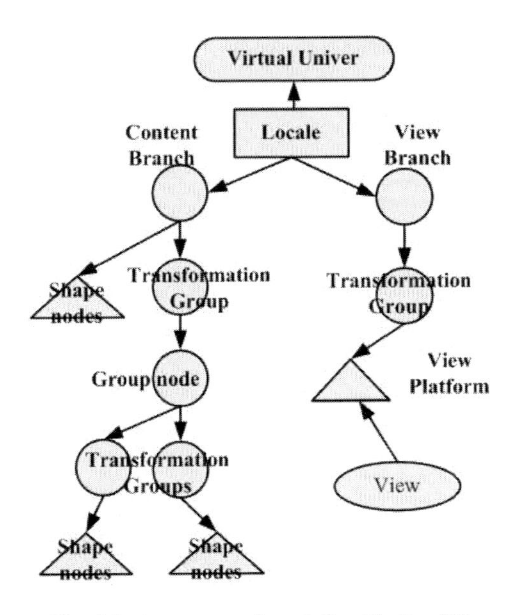

Fig. 1.9. A scene graph as defined in Java3D

The different terminologies employed in graph theory, in the object-oriented approaches and in scene or document representations such as VRML or XML are described in the following table.

Table 1.3. Terminologies employed in scene representation as graph or network

Graph Theory	Semantic Networks (OO Approach)	Scene Graph (VRML, Open-Inventor)	Structure Diagrams (XML)
Graph G= (N, E)	Schema (associations)	Scene Graph (DAG)	Structure (Tree)
Nodes	Classes (Entities) Objects (Instances)	Classes Nodes	Elements
Edges	Relations	Relations	Relations
Associated values	Attributes, Properties	Camps	Attributes

1.3.2 Conceptual Scene Models – Generic Models

In this section we focus in the representation of generic models based in semantic networks and the realisation of scene database schema for connectional models.

Semantic Network-based Conceptual Representations

Scene Descriptions in Restricted Natural Language: Systems using a restricted form of natural language for the description of spatial scenes at a high level have appeared during the 1990's [16]. The process of understanding requires the integration of artificial intelligence techniques (natural language understanding, spatial reasoning, rule-based reasoning and common-sense reasoning). The scene representation is accomplished through a description which is subsequently transformed into a *semantic network of spatial relations.* Figure 1.10 demonstrates such an example.

The workstation is on the desk
The books are on the desk
The pencil is on the right of the workstation
The pen is near the pencil

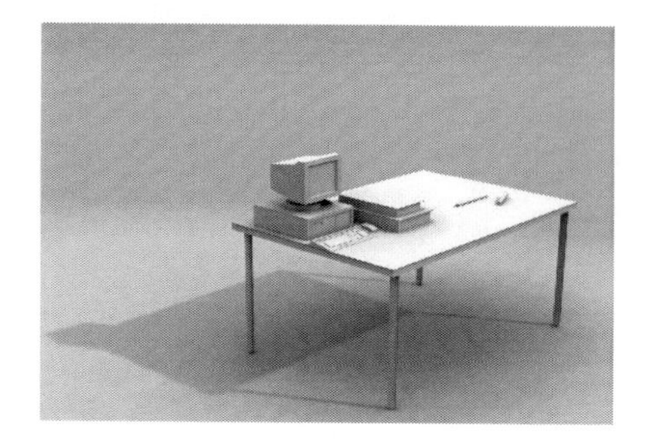

Fig. 1.10. From text to image

Declarative modelling by hierarchical decomposition: DMHD employs a technique of hierarchical decomposition which allows the construction of a hierarchical description based on a scene [27]. The concept of hierarchical description was introduced in order to facilitate the description of the scene properties at different levels of detail.

In case the designer is able to describe an *elementary scene*, the latter is described through a number of property constraints, which may be *size properties* (inter-dimensions) or *form properties*. Otherwise, the scene is partially described through properties that are easy to be described and, subsequently, it is decomposed into a number of *sub-scenes* and the same process is applied to each one of them. In order to express the relations among these sub-scenes, *placement properties* and size properties (inter-scene) are used.

This type of representation is demonstrated in the following classic example of **residence with a garage**. The initial description is the one closer to natural language. This formulation of description is supposed to be the one expressed by the user himself/herself, as shown in Figure 1.11.

The Residence consists of a House and a Garage;
The House is attached to the left of the Garage;
The House is higher than wide;
The House is higher than long;
The Garage is rounded on top at 80%;
The House consists of Walls and a Roof;
The Roof is placed on the Walls;
The Roof is rounded on top at 70%;

Fig. 1.11. Textual declarative description by hierarchical decomposition

In the following, *Spatial Relations Graphs* and *Decomposition Trees* are used for the representation of certain aspects of the aforementioned examples.

The transformation of the spatial relations into a semantic network [16], based on the formal language representation of the NALIG type appears in Fig. 1.12.

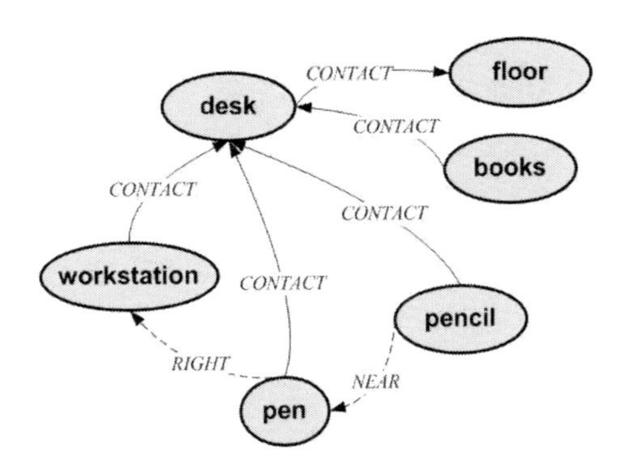

Fig. 1.12. Semantic network of spatial relations

Figure 1.13 presents the model of a scene description under the form of a hierarchical decomposition tree.

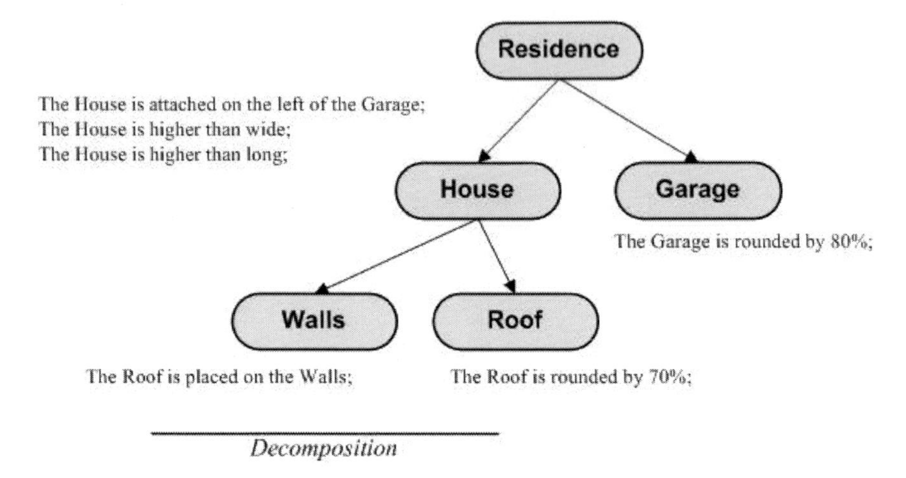

Fig. 1.13. The hierarchical decomposition tree

In another system ICADS [9] the scene is represented as a set of objects which are defined as a class hierarchy. The "part-of" relations are represented by a Spatial Relations Graph that contains as nodes instances of the classes of the object knowledge base. The type of information stored contains geometric data. The spatial attributes are extracted from the geometric data of the basic level provided by the geometric modeller of input.

The current spatial representations are mainly obtained from the area of natural language understanding, such as ON, UNDER, IN FRONT OF, …, FAR, NEAR, etc. Moreover, each spatial reference may be accompanied by a fuzzy qualifier represented by a prefix "-", "+" or "++". The spatial relations of ICADS are divided into multiple levels according to the decomposition hierarchy. This division technique is analogous to *partitioned semantic networks*.

1.3.3 Scene Conceptual Modelling in MultiCAD

The core of the system responsible for the representation and the storage of all information concerning the scenes in the context of the CADD systems we propose rely on conceptual representations of scenes which are compatible with the methodology of declarative modelling by hierarchical decomposition. We utilize the example of *residence with garage* in order to study the structure of a CADD system for conceptual scene modelling named MultiCAD.

As we have already discussed the Internal Declarative Representation (formal language based form of this model named IDR) [27] contains three types of elements, the *scenes* and *sub-scenes,* the *properties of a scene* and the *inter-scene properties*.

1. The **scenes** and **sub-scenes** are the objects produced by the decomposition or the synthesis of a scene.
2. The **properties of a scene** express the particularities of an object, for example the size properties (inter-dimensions) such as *higher than wide, as high as long,* etc.

or form properties (oblong, very rounded, etc.). In the latter case we also have parameterisation of the values (rounded = 80%).

3. The **inter-scene properties** express the relations among scenes, for example the placement properties (placed under, attached to the left, etc.) or the inter-scene size properties such as A higher than B, etc.

Using the aforementioned elements of representation, we may form two equivalent models for the residence example, as shown in Figure 1.14:

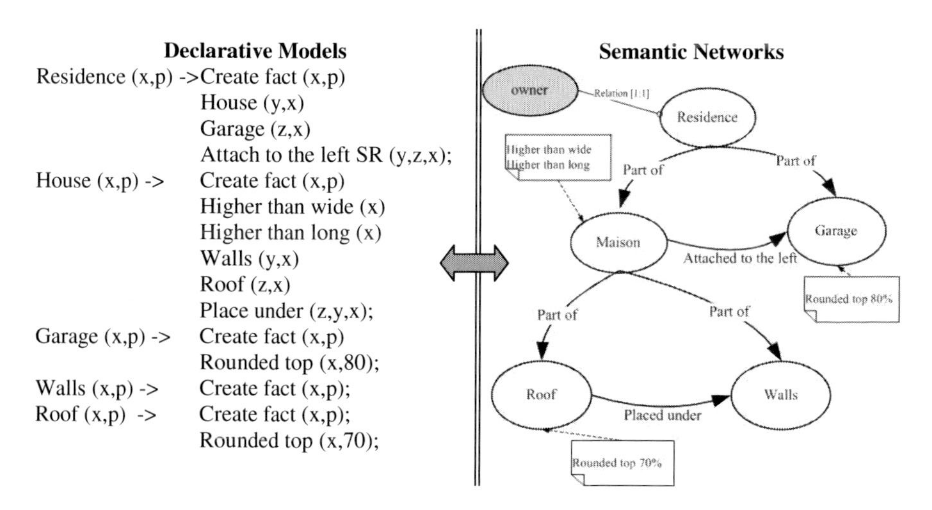

Fig. 1.14. Equivalence between declarative models and semantic networks

This equivalence of IDR and semantic network representations allows us to use semantic networks as the basis for scene modelling in MultiCAD. It also allows us to mutually transform these representations in the context of our system.

Scene Modelling Using Enhanced ER Models

The models based on *semantic networks*, as has been discussed, attempt to include both a network structure comprising nodes and edges as well as semantic constructions allowing a sense of the manipulated objects and their interconnection. The *Entity-Relationship model* makes a clear distinction between the objects identified in the model (the entities) and the links among these objects (the relations). It is, therefore, considered as an extension to relational models. The majority of the constraints are imposed by the user and are no longer connected to the model. The *enhanced ER model* augments to the initial entity-relationship model mainly with the concepts of aggregation and generalisation and offers the wealth of relations of the semantic models and the object models in a single model.

The **enhanced ER scene model** is a notion introduced in our research as an alternative representation of generic scene models to the text based forms. This notion combines the definition of a *general enhanced ER model* with certain notions of *scene graphs* presented in a previous section.

The Enhanced ER Scene Model or MultiCAD-Scene Model is a representation based on semantic networks, applying the general enhanced ER model and equivalent to the Internal Declarative Representations (IDR).

The MultiCAD-scene models are based on the definition of a Conceptual Scene Modelling Framework (CSMF) used to represent scenes in different components of an information system.

The conceptual model of a scene corresponds, by definition, to a set of geometric models of 3D objects. A geometric model of a 3D object yields a set of screen images corresponding to the different points of view of this concrete 3D object. Hence, a conceptual scene model is *a class of geometric models* (obtained by generation) and each geometric model corresponds to *a class of images* (points of view). A scene class could contain, in the extreme case, one and only one geometric model of a scene. This situation could be the result of constraints imposed by the combination of scene properties or by the (geometric) nature of certain properties, a case we should not exclude.

This perspective offers a unique view of the space of scene models where a geometric model (of declarative type) is one of the special cases of the generic model, abstract or conceptual. Our EER model must, therefore, be able to express all possible cases.

The scenes may also be simple, indivisible, the final elements of a decomposition or complex, consisting of a set of simple scenes (or components) following a decomposition hierarchy.

Another distinction is that of *visible* and *invisible* objects. By definition, the objects participating in a scene model have a physical, material nature. We have cases where entities affecting the elements of a scene in one or another way are not visualised in the scene's visual representation because this would not make sense. In a similar fashion, there are relations that affect the visual aspect of the scene and relations that do not, hence, we adopt the same distinction for *visible* and *invisible relations.* In our residence example, we may add as entities the functions of the pieces or their principal users. This distinction leads us to the transition from the EER Scene model to a regular EER model. The described property demonstrates the ability of the EER scene model to easily integrate non-geometric information.

An initial approach to the 3D-universe representation using the *EER Scene meta-model* which is at the basis of the framework for the conceptual scene modelling, is presented in Figure 1.15. The framework employs specific entities of two types, *Objects* and *Relations*. The 3D universe may, therefore, be described, using the entities of the model, as follows:

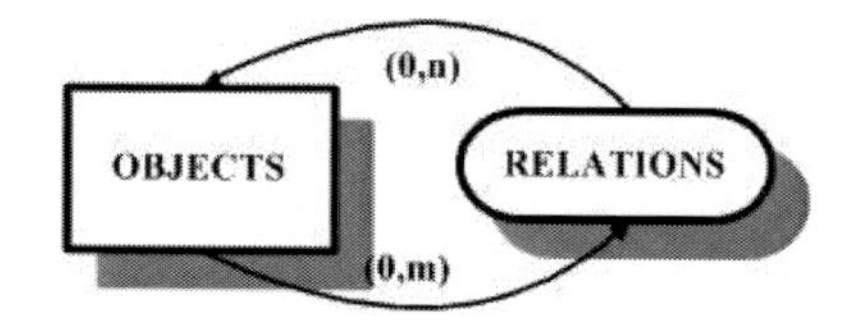

Fig. 1.15. Meta-model of 3D-universe representation

A scene description, under CSMF, consists of the elaboration of a specific conceptual model. Every model includes entities, their types of properties and characteristics and the multiplicity of their instance. A formal definition is necessary in order to render this representation more efficient and operational. In the following section we are going to define the elements of an EER scene model.

The entity types of the CSMF form a class hierarchy as shown in Fig. 1.16.

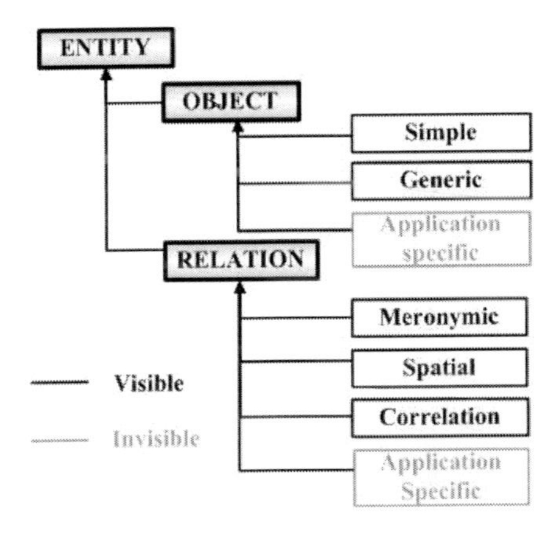

Fig. 1.16. Basic hierarchy for entity classes in CSMF

The hierarchy of entity classes of CSMF makes possible the *division of the meta-model* of representation of 3D-universes to allow the development of partial models destined for specific functions (such as for separately editing synthesis hierarchies or for analysing by level the spatial relations).

For example, according to the relation types described in the class hierarchy of Figure 1.17) we may decompose the general representation of the 3D-universe in four distinct representations: The *scene decomposition hierarchy,* the *spatial arrangement relations* of a collective object, the *application-oriented relations* of the scene (invisible), the *correlation relations* among objects of the same scene or a combination of these (Figure 1.12).

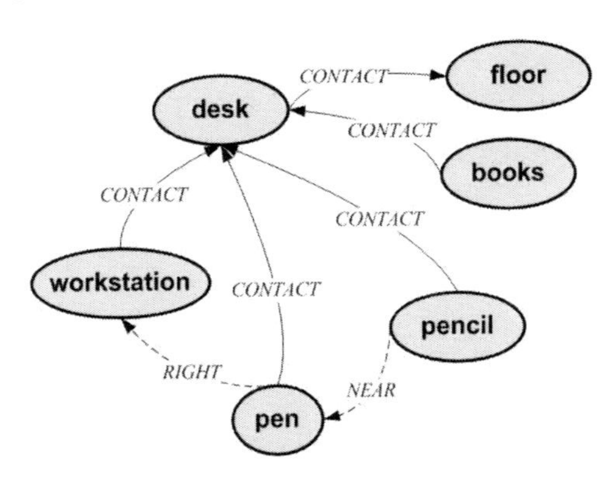

Fig. 1.17. The four distinct representations

Based on the definition of the elements of the Enhanced ER scene model, we have chosen a *EER scene graphical representation* and notation and we have applied this notation to our example of a residence with a garage. We have simply added two new elements to the example in order to enable the utilisation of the entire notation set. The first is the inter-scene property "the house is higher than the Garage" and the second is the invisible object representing the owner of the residence. The model for the residence example is shown in Fig. 1.18.

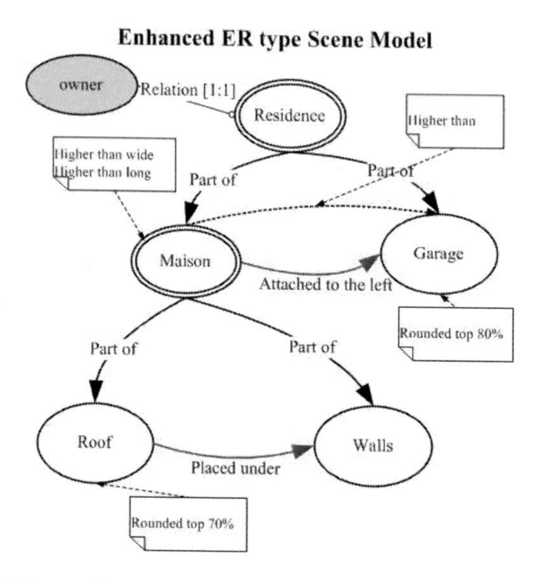

Fig. 1.18. The graphical version of a scene model of the Enhanced ER type

Organization of the scene models set in MultiCAD

After the complete definition of the elements of a scene model, we may propose the complete meta-model of the Conceptual Scene Modelling Framework (CSMF). These essential elements lay the foundations for our view of information and knowledge representation in MultiCAD. The data representation meta-model of the MultiCAD-scene model follows the norms of the object-relational models and it is presented as shown in Figure 1.19.

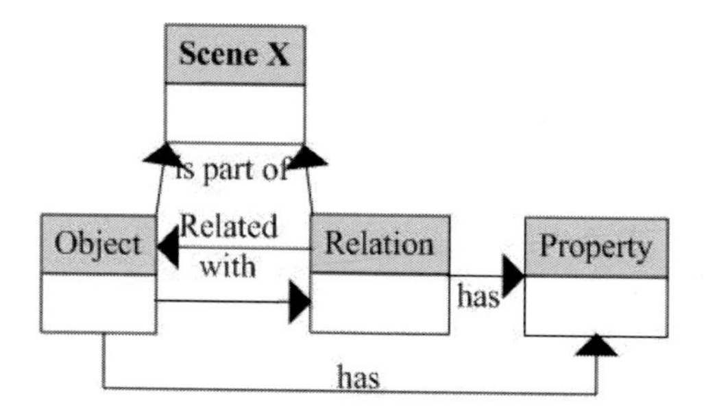

Fig. 1.19. Conceptual Scene meta-model

The organization schema of a database containing the entire set of scene models, i.e. conceptual, geometric and images directly yields the scene meta-model which is the centre of this organization (Figure 1.20).

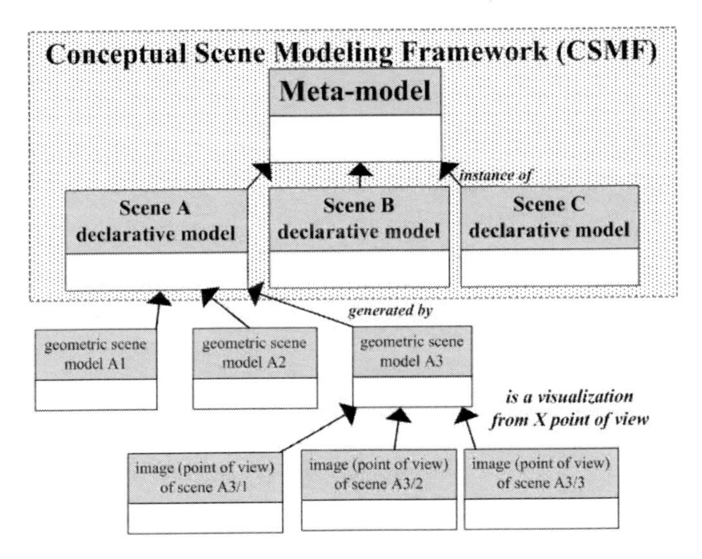

Fig. 1.20. Organisation Framework for CSMF scene models

1.4 Software Architectures for Declarative Design Support

In this section we will confront questions about the structure of software that supports declarative design. The goal is to make clear our point of view about the notions of *software architecture* and *framework architecture.* We also examine the passage from a framework architecture to the concrete MultiCAD software through an *architecture based development.*

We will define, at this point, the **MultiCAD architecture**. The domain of application in our case is the declarative design system, its *functional framework* and its *resources.*

In the first part, we begin with a short resume of the objectives and the constraints which are imposed in the development of this architecture-framework. Subsequently this part is related to the definition of the *functions* of MultiCAD software, in relation with (a) functional framework of the declarative design system, (b) functions relative to the management and exploitation of the *declarative design system's resources.* These functions are regrouped as big sets under the title *working environments.*

In the second part we define the *projection* of these sets of MultiCAD functions over a structure of appropriate layers, compatible with the characteristics of this family of software. Then we define the modules that correspond to each layer, which in turn they include the functionally homogenised components. At the end of this part we present a general view of the MultiCAD architecture.

1.4.1 MultiCAD: Objectives, Constraints and Functional Choices

The goals of the MultiCAD architecture are to ensure the development of information systems that will permit the evolution of the declarative modelling towards autonomous real design, supported by appropriate software tools and by a dedicated information and knowledge system, featuring instruments adapted to activities relative to the use of such resources [6, 10]. They are also aimed at the preparation of a necessary environment for the integration of the design process and the possible use of multiple design paradigms, and the presence of necessary conditions for the development of co-operative software systems, which will cover the whole design cycle [18, 19].

The nature of the MultiCAD-type applications is resulted not only by the goals and the functional framework, but also by technological choices and constraints established for the corresponding information system in its entirety.

The MultiCAD-type software are specialised applications like computer aided design systems, supporting the design of physical objects and consequently the development of models of these objects, with the use of image synthesis techniques within a three-dimensional space. The technology that enables the manipulation and utilisation of knowledge bases classify MultiCAD in the category of intelligent systems, and in the category of knowledge-based systems [30, 31]. One of the objectives is to guarantee the coherence of the different parts of the software, and to provide an integrated system. Also, MultiCAD must ensure the management and preservation of multimedia information that they constitute crucial resources for the design process. The human-machine interaction modes, adapted by the declarative design process as well as by the conventional CAD software systems, indicate the support of a strong interactivity and a high quality of graphic interfaces. The last constraint concerns the distribution of MultiCAD software modules, and the possibilities of parallel processing. The system architecture must facilitate the distribution of software modules, the information and the knowledge of the system. With regard to parallel processing, it is addressed rather to the heavy processing, such as a generation of solutions [5]. A modular architecture could facilitate the integration of these engines within a declarative design support system.

The functional choices of MultiCAD software yield as results models of design systems and models of data and corresponding knowledge.

The *set of the supported functions* by MultiCAD follows the functional decomposition in three groups relative to the previously presented process of design. This group of functions corresponds to distinct working environments which we propose:

1. **Project Planning** (project planning and management, modelling relative to ideas, needs and functions)
2. **Design** (total design, declarative cycle, case manipulation, geometric modelling)
3. **Management of resources** (Knowledge, Information, Cases)

We have to notice that there exist functions which are presented to the user in the context of many environments. Evidently, these functions all invoke the same inferior layers' components independently of their repetition.

1.4.2 Definition of MultiCAD Framework-Architecture

The structure of the MultiCAD system software is based on the common architectural paradigm of "layered architecture". This paradigm permits the definition of components' families to hierarchically organised software layers. These components also belong to the different functional groups that we have already proposed. The components of every layer provide services to components of the superior layer, and they are also served as clients by the inferior layer.

The layers refer to the architectural models of the system which deal with the resources management of important information. The three typical elements of such systems are: (a) resources of information and knowledge themselves; (b) components corresponding to specific treatments; and (c) infrastructure of the management of resources bases and communication. In these elements we have to add the interactive dimensions that are represented by the user interface component.

We may now define a model of an architecture that is based upon three layers. In this model we project the groups of functions corresponding to preliminary phases, in the design and management of resources. The three layers (shown in Figure 1.21) are the following:

- An **Interface layer**, which is decomposed in three elements, (a) a mechanism that controls the entire application, i.e. all the working environments, (b) a sub-layer of presentation, (c) a sub-layer of dialogue. The interface layer is responsible for the intelligent visualisation, for interfaces that enable model and description creation and editing, for the control of the generation process, for the request formulation, for information navigation, for information acquisition and editing.
- A **Process layer**, which includes specialised components for, (a) necessary processing for supporting the modellers of the declarative models, of geometric models, or other kinds of models, (b) for horizontal and vertical transformation in the abstraction axis of models of the entities processed by the system.
- An **Information and knowledge management layer** is used for structuring, management, searching and exploitation of the different databases (scenes, cases, concepts, knowledge, documents, multimedia). Additionally it is used for the different types of communication between other sites or processes.

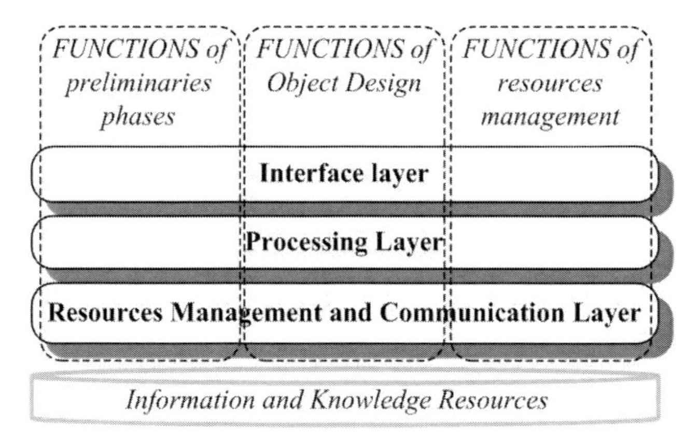

Fig. 1.21. A basic three-layered architecture

The last elements of the model are the resources of information and knowledge themselves which are stored in the databases according to the MultiCAD platform.

The Resources of Information and Knowledge

Within the MultiCAD system framework many types of information and knowledge coexist, like scene models, general information, relative or associated with a scene or a project, multimedia documents, and declarative knowledge and domain concepts (Fig. 1.22.).

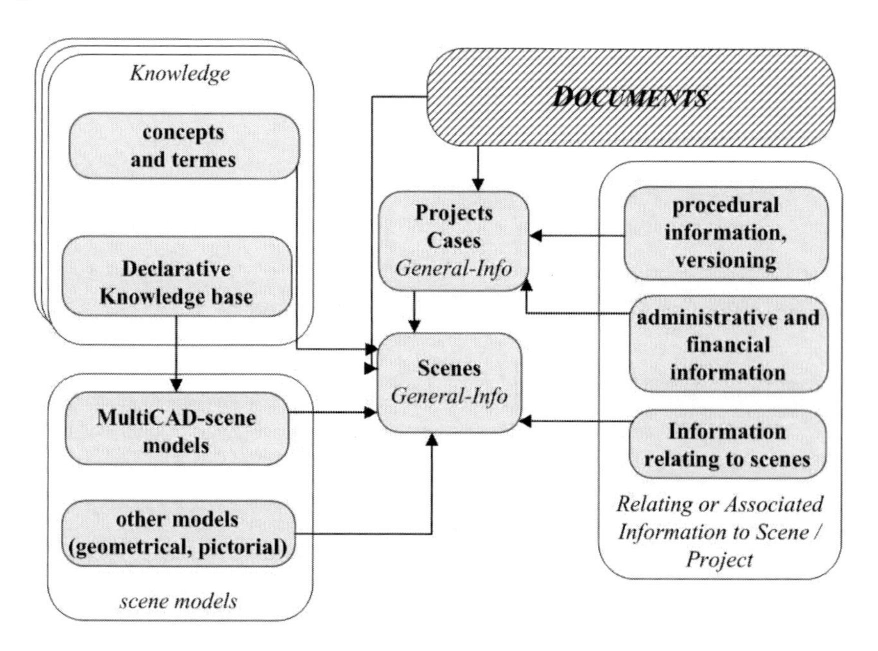

Fig. 1.22. Information and knowledge resources

Many other parameters relative to different processing activities, such as the designer's attitude are equally stored. In particular:

- **Scene models:** Abstract (declarative) scene models, geometric models and images.
- **Information relative to a scene, or associated with a scene or a project:** Information relative to scenes, administrative and financial information, information about the procedures, versions of the scene models (products of different design phases).
- **General information about scenes and projects:** General characteristics of a scene or a project.
- **Documents:** Multimedia documents connected to scenes or to projects.
- **Knowledge:** Models of declarative knowledge, structures of concepts and terms from different specific domains of know-how.

The storage system is organised according to the MultiCAD platform.

Overall view of the MultiCAD architecture

In fig. 1.23 we present an overall view of the MultiCAD architecture, where they are presented the modules of the components and the interactions with the designer and the operating system.

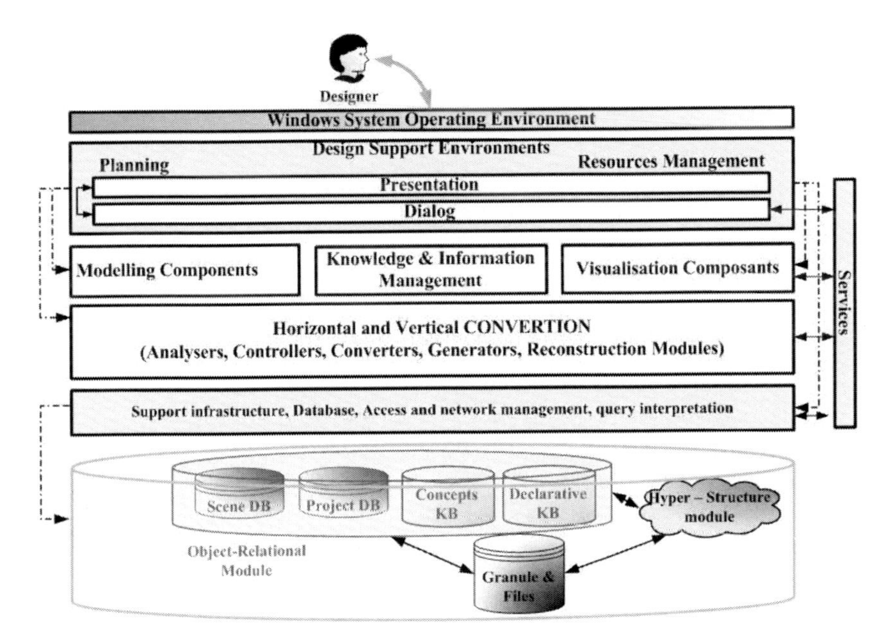

Fig. 1.23. The MultiCAD architecture, overall view

1.5 Conclusion

In this chapter we have presented and analysed the principal key points of an intelligent information system dedicated to scene modelling using declarative design methodology.

The general approach that was introduced has utilised the analysis of a declarative design process mechanisms as instrument to propound a functional framework for a wide range of information systems. The analysis of the particularity of the informational and knowledge contents of this kind of systems was the basis for a set of propositions in order to organise the appropriate repositories. Special emphasis was put on two key dimensions (a) the scene model representation and (b) the semantic organisation of the adequate abstraction hierarchy of those models. The software architecture that was formed integrates the functional and informational aspect of those systems into a coherent set.

This software architecture was used in order to develop an experimental implementation of a system called MultiCAD. This system using a constraint resolution engine offered a satisfactory platform to demonstrate the applicability of the aforementioned principles.

The basic core of this platform was proved able to integrate the extensions of the system e.g. machine learning mechanisms, personalisation modules, reverse engineering process capabilities, styling modules and collaborative environments [3, 4, 13, 15, 17, 22].

We also find that there is a significant opportunity to improve conventional design support systems by adopting this coherent system architecture, reengineering limited number of modules and evolve toward an intelligent scene modelling information system able to cover advanced design needs.

References

1. Andersson, K.: A Design Process Model for Multiview Behavior Simulations of Complex Products. In: Proceedings of DETC 1999, ASME Design Engineering Technical Conferences, Las Vegas, Nevada, September 12-15 1999)
2. Arciszewski, T.: Design Theory and Methodology in Eastern Europe, ASME. In: DTM 1990, Chicago, IL, vol. DE-27, pp. 209–217 (1990)
3. Bardis, G.: Machine Learning and Decision Support for Declarative Scene Modelling / Apprentissage et aide à la décision pour la modélisation déclarative de scènes (bilingual), Thèse de Doctorat, Université de Limoges, France (2006)
4. Bardis, G., Miaoulis, G., Plemenos, D.: User Profiling from Imbalanced Data in a Declarative Modelling Environment. In: Techniques for Computer Graphics, pp. 123–140. Springer, Heidelberg (2008)
5. Bonnefoi, P.-F.: Techniques de satisfaction de contraintes de la modélisation déclarative. Application à la génération concurrente de scènes Thèse de doctorat, Université de Limoges (1999)
6. Borst, W.N.: Construction of Engineering Ontologies for Knowledge Sharing and Reuse, Ph.D. Thesis, SIKS, Dutch Graduate School for Information and Knowledge Systems, University of Twente, the Netherlands (1997)
7. Braha, D., Maimon, O.: The Design Process, Properties, Paradigms, and Structure. IEEE Trans. on systems, man. and cybernetics - part A: systems and humans 27(2), 146–166 (1997)
8. Chauvat, D.: The VoluFormes Project: An Example of Declarative Modelling with Spatial Control, PhD Thesis, Nantes, France (1994)
9. Chun, H.W., Ming-Kit Lai, E.: Intelligent Critic System for Architectural Design. IEEE Trans On Knowledge and Data Engineering 9(4), 625–639 (1997)
10. Coyne, R.D., Rosenman, M.A., Radford, A.D., Balachandran, M., Gero, J.S.: Knowledge-based Design Systems. Addison-Wesley, Reading (1990)
11. Tate, D.: A Roadmap For Decomposition: Activities, Theories, And Tools For System Design PhD Dissertation, Department of Mechanical Engineering. MIT (1999)
12. Guillaume, D.: Étude et réalisation de techniques de prise de connaissance de scènes tridimensionnelles Thèse de Doctorat, Université de Limoges (2001)
13. Dragonas, J.: Collaborative Declarative Modelling / Modelisation Declarative Collaborative (bilingual), Thèse de Doctorat, Université de Limoges, France (2006)
14. Evbuomwan, N.F.O., Sivaloganathan, S., Jebb, A.: A Survey of Design Philosophies, Models, Methods and System. Proceedings IMechE Part B: Journal of Engineering Manufacture 210(B4), 301–320 (1996)
15. Fribault, P.: Modelisation Declarative d'Espaces Habitable (in French), Thèse de Doctorat, Université de Limoges, France (2003)

16. Giunchiglia, E., Armando, A., Traverso, P., Cimatti, A.: Visual Representation of Natural Language Scene Descriptions. IEEE Trans. On Systems, Man, And Cybernetics-Part B: Cybernetics 26(4), 575–588 (1996)
17. Golfinopoulos, V.: Study and Implementation of a Knowledge-based Reverse Engineering System for Declarative Scene Modelling / Étude et réalisation d´un système de rétro-conception basé sur la connaissance pour la modélisation déclarative de scènes (bilingual), Thèse de Doctorat, Université de Limoges, France (2006)
18. Kochhar, S.: Cooperative Computer-Aided Design PhD dissertation Harvard University (1990)
19. Kochhar, S.: CCAD: A Paradigm for Human-Computer Cooperation in Design. IEEE Computer Graphics and Applications (May 1994)
20. Lucas, M., Desmontils, E.: Les modeleurs déclaratifs Paper in Revue de CFAO (1996)
21. Lucas, M., Martin, D., Martin, P., Plemenos, D.: The ExploFormes project: Some Steps Towards Declarative Modelling of Forms. In: AFCET-GROPLAN Conference, Strasbourg, France, vol. (67), pp. 35–49. BIGRE (1990) (in French)
22. Makris, D.: Study and Realisation of a Declarative System for Modelling and Generation of Style with Genetic Algorithms: Application in Architectural Design / Etude et réalisation d'un système déclaratif de modélisation et de génération de styles par algorithmes génétiques (bilingual), Thèse de Doctorat, Université de Limoges, France (2005)
23. Martin, D., Martin, P.: PolyFormes: Software for the Declarative Modelling of Polyhedra. The Visual Computer 15, 55–76 (1999)
24. Miaoulis, G.: Contribution à l'étude des Systèmes d'Information Multimédia et Intelligent dédiés à la Conception Déclarative Assistée par l'Ordinateur – Le projet MultiCAD, Thèse de Doctorat, Université de Limoges, France (2002)
25. Miaoulis, G., Plemenos, D., Magos, D.: The MultiCAD Project: toward an intelligent multimedia information system for CAD. In: Proceedings of WSCG 2000 the 8-th Int. Conf.in Central Europe on Computer Graphics, Visualization and Interactive Digital Media 2000, Plzen, Czech, February 7-11 (2000)
26. Miaoulis, G., Plemenos, D., Skourlas, C.: MultiCAD Database: Toward a unified data and knowledge representation for database scene modelling. In: 3rd 3IA International Conference on Computer Graphics and Artificial Intelligence, Limoges, France, May 3-4, 2000, pp. 147–152 (2000)
27. Plemenos, D.: Declarative modelling by hierarchical decomposition. The actual state of the MultiFormes project. In: Communication in International Conference GraphiCon 1995, St Pe-tersburg, Russia (1995)
28. Plemenos, D., Miaoulis, G., Vassilas, N.: Machine learning for a General Purpose Declarative Scene Modeller. In: International Conference GraphiCon 2002, Nizhny Novgorod (2002)
29. Russel, S., Norwig, P.: Artificial Intelligence – A Modern Approach, 2nd edn. Prentice-Hall, Englewood Cliffs (2002)
30. Thornton, A.C., Johnson, A.: Constraint specification and satisfaction in embodiment design. In: Proceedings of ICED 1993, The Hague, vol. 3, pp. 1319–1326 (1993)
31. Tong, C., Sriram, R.D. (eds.): Artificial Intelligence in Engineering Design. Academic Press, London (1992)
32. Wood III, W.H., Agogino, A.M.: Case-based conceptual design information server for concurrent engineering in Computer-Aided Design 28(5), 361–369 (1996)
33. Zeisel, J.: Inquiry by Design Monterey. Brooks/Cole, CA (1981)

2

Declarative Modeling in Computer Graphics

Dimitri Plemenos

University of Limoges
plemenos@numericable.com, plemenos@unilim.fr
http://msi.unilim.fr/~plemenos

Abstract. A review of declarative scene modeling techniques is presented in this paper. After a definition of the purpose of declarative modeling, some existing declarative modelers are classified according to the manner to manage imprecision in scene description. Then, a concrete application is described, using constraint satisfaction techniques in the framework of Declarative Modeling by Hierarchical Decomposition (DMHD). Machine-learning techniques, applied to DMHD are presented as well. The aim of this chapter is to present declarative scene modeling and to show its importance for a really computer aided design in image-based information systems, as well as some drawbacks of this modeling technique. Some suggestions for possible future extensions of declarative modeling are also given.

Keywords: Declarative modeling, Geometric modeling, Constraint Satisfaction Problem, Machine-learning, Scene understanding.

2.1 Introduction

Scene modeling is a very difficult task in computer graphics, as traditional geometric modelers are not well adapted to computer aided design. With most of the current modeling tools the user must have quite precise idea of the scene to design before using a modeler to achieve the modeling task. In this manner, the design is not really a computer aided one, because the main creative ideas have been elaborated without any help of the modeler.

The problem with most of the current scene modelers is that they need, very soon during the designing process, low-level details which are not important in the creative phase of design. This is due to the lack of levels of abstraction allowing the user to validate general ideas before resolve low-level problems.

If the initial very general idea of the designer is, for example, to design a scene comporting a house, a swimming tool in front of the house and a tree on one side of it, this idea may be realized in many different manners. As the modeler does not offer the user an easy manner to try and test different manners to realize the initial mental idea, he (she) generally tries a small number of possible solutions and chooses the best one. In this manner, the user may lack very interesting possible solutions.

Declarative modeling tries to give intuitive solutions to this kind of problem by using Artificial Intelligence techniques which allow the user to describe high level properties of a scene and the modeler to give all the solutions corresponding to imprecise properties.

G. Miaoulis and D. Plemenos (Eds.): Intel. Scene Mod. Information Systems, SCI 181, pp. 29–57.
springerlink.com

The remaining of this chapter is organized as follows: In section 2, a brief description of declarative modeling in computer graphics will be done. In section 3, the main imprecision management modes, used by declarative modelers during the scene generation process, are presented. In section 4, some important realizations in the area of declarative modeling will be presented, classified according to the manner they use to manage imprecise descriptions. In section 5 we will present techniques allowing to understand and to evaluate the pertinence of a solution obtained by a declarative modeler, while in section 6 declarative scene modeling based on CSP resolution techniques will be presented. Examples of machine-learning techniques applied to declarative scene modeling, in order to improve obtained results, will be given in section 7. Advantages and drawbacks of declarative scene modeling will be discussed in section 8. In section 9 we will discuss future issues of declarative scene modeling before concluding in section 10.

2.2 What Is Declarative Scene Modeling

Declarative modeling [1, 2, 3, 5, 38] in computer graphics is a very powerful technique allowing to describe the scene to be designed in an intuitive manner, by only giving some expected properties of the scene and letting the modeler find solutions, if any, verifying these properties.

As the user may describe a scene in an intuitive manner, using common expressions, the described properties are often imprecise. For example, the user can tell the modeler that "the scene A must be put on the left of scene B". There exist several possibilities to put a scene on the left of another one. Another kind of imprecision is due to the fact that the designer does not know the exact property his (her) scene has to satisfy and expects some proposals from the modeler. So, the user can indicate that "the house A must be near the house B" without giving any other precision. Due to this lack of precision, declarative modeling is generally a time consuming scene modeling technique.

There exist two kinds of geometric modelers, general purpose modelers, allowing to design almost everything, and specialized (or dedicated) modelers, offering high level modeling for limited specific modeling areas. In the same manner, there exist two families of declarative modelers: general purpose modelers, covering a large set of possible applications, and dedicated modelers, covering a specific area (architectural design, mechanical design, …).

The principle of dedicated modeling is to define a declarative modeler each time it is necessary for a well delimited modeling area (see Fig. 2.1). Thus, PolyFormes [6,7] is a declarative modeler designed to generate regular or semi-regular polyhedrons.

The main advantage of the dedicated declarative modelers is efficiency because their solution generation engine can be well adapted to the properties of the specific modeling area covered by the modeler. On the other hand, it is difficult for such a modeler to evolve in order to be able to process another specific modeling area.

The aim of the general purpose modelers is generality. These modelers include a solution generation engine which can process several kinds of properties, together with a reduced set of pre-defined properties, as general as possible. General purpose declarative modelers could normally be specialized in a specific modeling area by

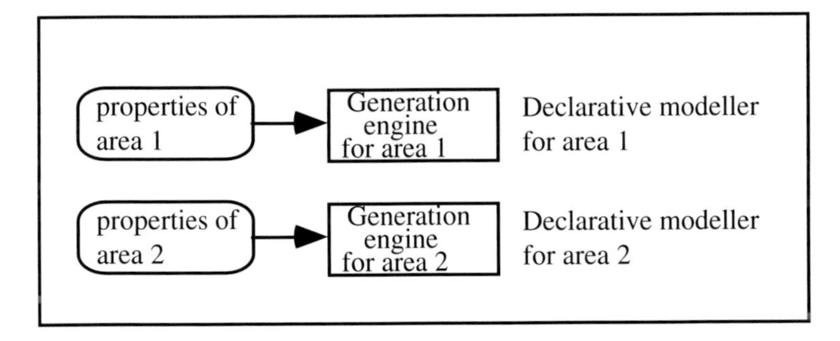

Fig. 2.1. Dedicated declarative modelers

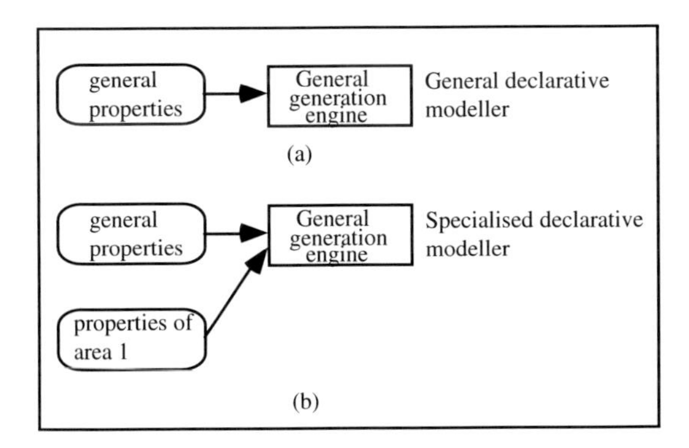

Fig. 2.2. General purpose declarative modeler (a) and its specialization (b)

adding to them new properties, corresponding to the specific modeling area we want to cover (see Fig. 2.2). In this sense, general purpose modelers can be seen as platforms to generate dedicated declarative modelers.

The main advantage of general purpose declarative modelers is generality which allows to specialize a modeler in a specific modeling area without having to modify its solution generation engine. On the other hand, general purpose modelers suffer from their lack of efficiency, because of the generality of the solution generation mechanism.

The declarative modeler MultiFormes [2 ,3, 4, 12, 13, 14, 34] is a general purpose declarative modeler.

It is generally admitted that the declarative modeling process is made of three phases: the *description* phase, where the designer describes the scene; the *scene generation* phase, where the modeler generates one or more scenes verifying the description; and the *scene understanding* phase, where the designer, or the modeler, tries to understand a generated scene in order to decide whether the proposed solution is a satisfactory one, or not.

2.3 Imprecision Management in Declarative Modelers

Declarative modeling tries to help the scene designer by allowing intuitive descriptions using a “language” close to the user’s one. This kind of description is very often imprecise and can produce many solutions. The modeler has to manage this imprecision in the scene generation phase. Two modes are used by declarative modelers to manage imprecision during the generation phase: *exploration* mode and *solution search* mode.

In exploration mode, the declarative modeler, starting from a user description, performs a full exploration of the solution space and gives the user all found solutions. This mode can be used when the designer has insufficient knowledge of a domain and wants to discover it by an exhaustive exploration or when the designer is looking for new ideas and hopes that the modeler could help him (her) by exploring a vague description. The use of imprecise properties increases the richness of the solution space and allows the user to obtain concrete answers for a vague mental image. So, the use of imprecise properties is very important for the designer. As the exploration mode is based on the use of imprecise properties, it is very important to have techniques to reduce exploration cost by reducing the number of useless tries during the solution search process [12, 13]. A problem with the exploration mode is that the use of general imprecise properties can produce a very important number of solutions and make very difficult the management of these solutions. Furthermore, some families of solutions can be of no interest for the designer and he (she) would like to avoid generation of such solutions in subsequent generations. As the modeler does not know the designer's preferences, interaction is necessary to learn it what kind of scenes are not interesting.

In solution search mode, the modeler generally generates only one solution. To do this, either it interprets the designer's description in a restrictive manner or it asks the designer to precise his (her) choice. So, the designer must have a relatively precise idea of the kind of scenes he (she) would like to obtain.

Declarative modelers working in exploration mode are, generally, able to work in solution search mode if the designer would like to obtain a solution immediately or very quickly from a description using less imprecise properties. As the semantic of a property is often ambiguous and several solutions not satisfactory for the user can be faced by the modeler, the designer must have the possibility to interactively intervene in order to guide the modeler in its search. So, if parts of a solution proposed by the modeler are close to the idea the designer has of the wished scene, the designer should be able to tell the modeler not to modify these parts in proposed solutions. This interaction decreases the solution space because, for a great number of scenes verifying the properties of the initial description, these parts will not satisfy the intuitive idea of the user and these scenes will be avoided.

2.4 A Classification of Declarative Scene Modelers

In this section will be presented some experimental declarative scene modelers developed since 1987, date of the beginning of declarative scene modeling. These modelers will be classified according to the mode - *exploration* mode or *solution search* mode - of imprecision management they can support during the scene generation phase.

2.4.1 Modelers Using Exploration Mode in Scene Generation

PolyFormes [6, 7]. The first experimental declarative modeler was developed in Nantes (France). Its first version was written in 1987 in Prolog. Other versions have

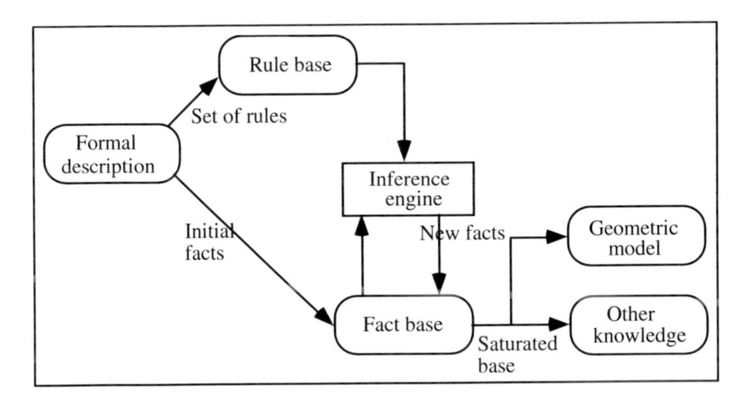

Fig. 2.3. The internal deduction mechanism of PolyFormes

been developed later in Pascal. The goal of the PolyFormes declarative modeler is to generate all regular and semi-regular polyhedrons, or a part of the whole, according to the user's request.

Requests may be more or less precise and are expressed using dialog boxes. This initial description is then translated in an internal model which will be used during the generation process. This internal model is a knowledge base, made of a rule base and a fact base.

The scene generation process, uses an inference engine which applies rules of the rule base to the facts of the fact base and creates new facts. A solution is obtained by saturation of the fact base. The whole modeler is an expert system on polyhedrons. When the initial description is imprecise, all the possible solutions are generated. Fig. 2.3 shows the internal deduction mechanism of PolyFormes.

In Fig. 2.4 one can see an example of polyhedron generated by the PolyFormes declarative modeler. PolyFormes is a dedicated declarative modeler as it is specialized in generation of polyhedrons.

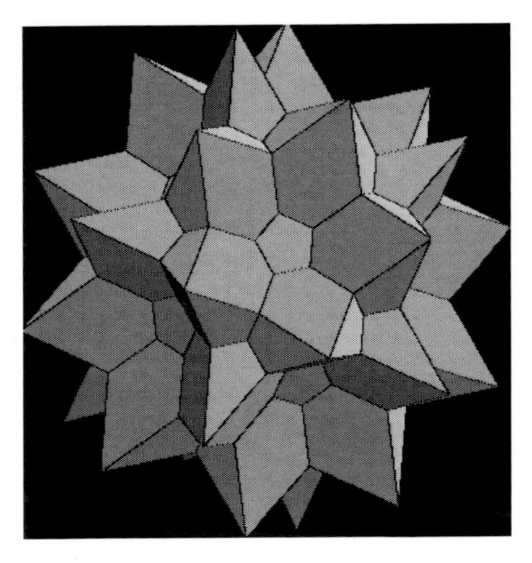

Fig. 2.4. Scene generated by PolyFormes

MultiFormes [2, 3, 4, 13, 14, 34]. The study of MultiFormes started in 1987 in Nantes (France) and its first version was available in 1991. Several other versions of the modeler have been developed later in Limoges (France). The purpose of this modeler was to be a general purpose declarative modeler, able to be specialized in any particular area.

MultiFormes is based on a new conception and modeling technique, declarative modeling by hierarchical decomposition (DMHD) [2, 3, 34]. The DMHD technique can be resumed as follows:

- If the current scene can be described using a small number of predefined high level properties, describe it.
- Otherwise, describe what is possible and then decompose the scene in a number of sub-scenes. Apply the DMHD technique to each sub-scene.

Descriptions in MultiFormes are expressed by means of dialog boxes allowing to represent a tree-like structure, to select nodes and to assign them properties. The initial description is then translated to an internal model to be used during the scene generation process.

In the first version of MultiFormes the internal model was a knowledge base made of a set of rules and a set of facts. In all the other versions of the modeler, the internal model is a set of arithmetic constraints on finite domains.

The scene generation process uses a constraint satisfaction engine which applies CSP (Computer Satisfaction Problem) techniques to generate all the solutions corresponding to a given description. A special form of primitive arithmetic constraints, CLP (FD) is used to improve the scene generation process.

The tree of the hierarchical description of a scene, used in the scene generation phase, allows scene generation in various levels of detail and reduction of the generation's cost. To do this, the modeler uses a bounding box for each node of the tree. This bounding box is the bounding box of the sub-scene represented by the sub-tree whose the current node is the root. All bounding boxes of the children nodes of a node are physically included in the bounding box of the parent node. This property permits to detect very soon branches of the generation tree which cannot be solutions. In Fig. 2.5, the spatial relation between the bounding boxes of a scene and its sub-scenes is shown.

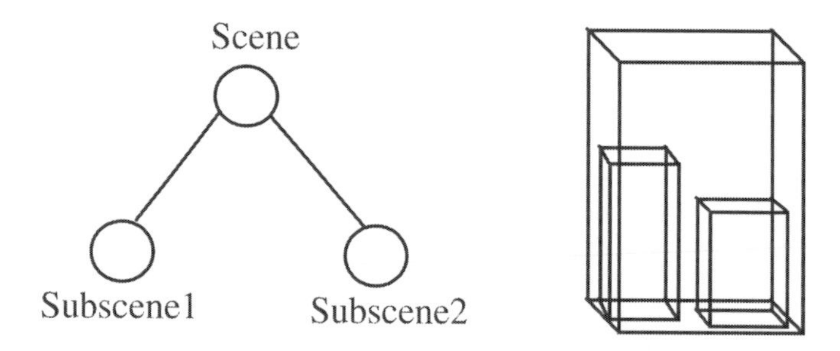

Fig. 2.5. The bounding boxes of the sub-scenes of a scene are inside the bounding box of the parent scene

Fig. 2.6. Cathedral off Le Dorat (France) designed by W. Ruchaud

MultiFormes is a general purpose declarative modeler which can be specialized by adding new predefined properties. The scene generation phase works in exploration mode, whereas it is possible to use solution search mode by means of user's interaction.

MultiFormes can also use a set of geometric constraints instead of arithmetic ones. This set contains constraints like "point P is in the box B" or "Points P1, P2, P3 are aligned". Satisfaction of this kind of constraints is computed using CSP-like techniques and allows more flexibility in creating scenes [35, 38].

An example of scene generated by geometric constraint satisfaction techniques can be seen in Fig. 2.6. Machine learning techniques based on neural networks have been implemented in MultiFormes. These techniques allow the modeler to select scenes close to the designer's desires in solution search mode, during the scene generation phase.

A version of MultiFormes was integrated in MultiCAD, an intelligent image-based information system [45]. Several research work is currently in progress on the MultiCAD system [41, 42, 43, 44] in order to get a more intelligent system, able to apply machine-learning techniques, to recognize architectural styles, to integrate traditional geometric modeling in the DMHD design process or to offer possibilities of cooperative scene modeling.

2.4.2 Modelers Using Solution Search Mode in Scene Generation

DE2MONS [24]. The DE2MONS declarative modeler is a general purpose modeler whose main properties are:

- A multi modal interface,
- A generation engine limited to the placement of objects,
- A constraint solver able to process dynamic and hierarchical constraints.

The modeler uses a multi modal interface allowing descriptions by means of the voice, the keyboard (natural language), a data glove or 3D captors informing the system of the user's position. The description is translated in an internal model made of linear constraints.

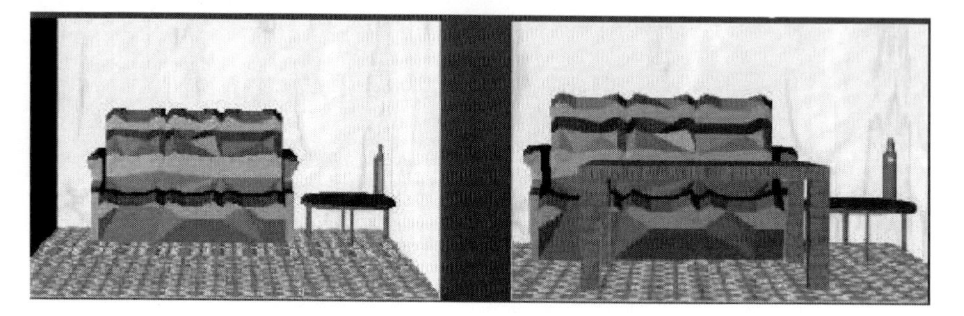

Fig. 2.7. Furniture pieces placement with DE2MONS

The generation engine of DE2MONS uses a linear constraint solver, ORANOS, able to process dynamic constraints (new constraints can be added during generation) and hierarchical constraints. Hierarchical constraints are constraints with priorities assigned by the user. Whenever there is no solution for a given description, constraints with low priority are released in order to always get a solution. The solver computes one solution for a given description.

Images in Fig. 2.7 show furniture pieces placements generated by the generation engine of DE2MONS.

CCAD [15, 16]. The Cooperative Computer Aided Design (CCAD) paradigm was introduced by S. Kochhar to facilitate the integration of *generative* and traditional modeling systems by allowing the designer to guide de generative system through successive rounds of automated geometric modeling.

The notion of generative modeling is very close to the notion of declarative modeling, as in both cases imprecise descriptions can generate many solutions. An experimental cooperative scene modeler was implemented for a generative system based on the formal language of schema grammars.

The CCAD framework is based on three main premises:

- A generative geometric modeling (GGM) system exists and can be used to generate a set of designs based on some designer-defined constraints or properties.
- The GGM system is supposed not to produce perfect designs, but rather it will be guided to search for better designs by the human designer.
- As the GGM system will produce a large set of designs, a specialized browsing system allows the designer to search the set of generated designs in a directed manner.

A typical modeling session using the CCAD system proceeds as follows:

- The designer uses the TGM system to generate a nascent design to be used in the first iteration of automated modeling.
- The designer then uses a dialog with the GGM system to define the constraints to be used during the generation process.
- The GGM system then instantiates all valid geometric designs. These designs are presented as icon-like buttons in a large screen area and the designer can get a large image of a design by clicking on the corresponding button.

- The designer then selects a set of promising designs using the browsing system.
- The selected designs are then returned to GGM system and the 4 four last steps are repeated until the desired design has been constructed.

Fig. 2.8 shows the architecture of the CCAD system. The CCAD paradigm has also been applied by D. Sellinger [17, 18] in a PhD thesis whose purpose was to integrate declarative and traditional geometric modeling.

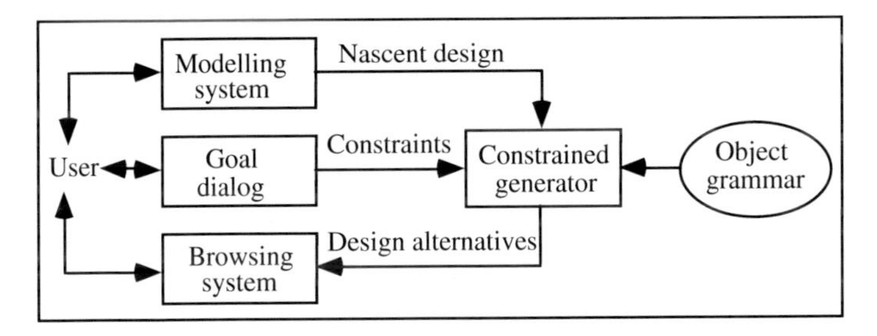

Fig. 2.8. The CCAD system architecture

VoluFormes [21]. VoluFormes is a *dedicated* declarative modeler allowing the user to quickly define boxes in the space whose purpose is to check the growth of forms. It is made of two modules:

- *Voluboites*, which allows to define boxes where the spatial control is performed.
- *Voluscenes*, which allows to use growth mechanisms applied to elementary germs and to create forms, taking into account the spatial control boxes.

Only Voluboites works in declarative manner. The positions of spatial control boxes are described during the description phase using a natural-like language. Description and generation work in incremental manner. Each box is placed in the 3D space and, if the user does not like the proposed box and placement, another solution

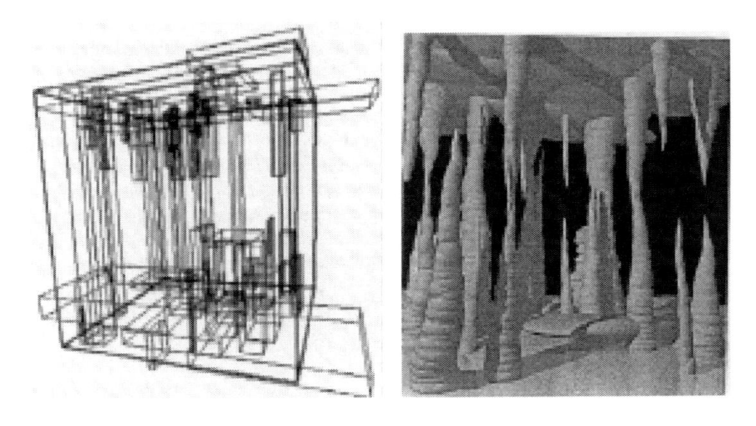

Fig. 2.9. Boxes arrangement and form growth with VoluFormes

can be given. Once the current box is placed in the space, the same process is applied to define the next one. The generation engine of Voluboites uses a CSP-like constraint satisfaction technique. On the left of Fig. 2.9, one can see a boxes arrangement obtained by Voluboites.

Voluscenes is used by the designer to put germs in the boxes and to select the growth mechanism, among a reduced number of predefined imperative mechanisms. On the right of Fig. 2.9 one can see an example of scene obtained by form growth.

2.4.3 Other Declarative or Declarative-Like Modelers

The declarative scene modelers presented in subsections 2.4.1 and 2.4.2 are the most important ones, as they summarize the main properties of declarative modeling in computer graphics. These modelers allow essentially to design the geometry of a scene. However, some other declarative scene modelers were developed, especially dedicated modelers, often permitting to design other, non geometrical aspects of a scene.

Pascale Fribault [31] developed Mosaïca, a dedicated scene modeler, based on a specialisation of the MultiFormes declarative modeler. Mosaïca is a modeler of livable areas, specialized in the design of pavilions.

Another scene modeler, QuickVox, developed by Mathieu Kergall [33] in Nantes, (in 1989) allows to design objects from their shadows on the three main planes defined by the three main axes of coordinates. The modeler is not really a declarative one but it may propose more than one solution from a given description. In Fig. 2.10 one can see an example of scene designed with QuickVox.

F. Poulet defined and developed a declarative modeler for designing megalithic sites [22].

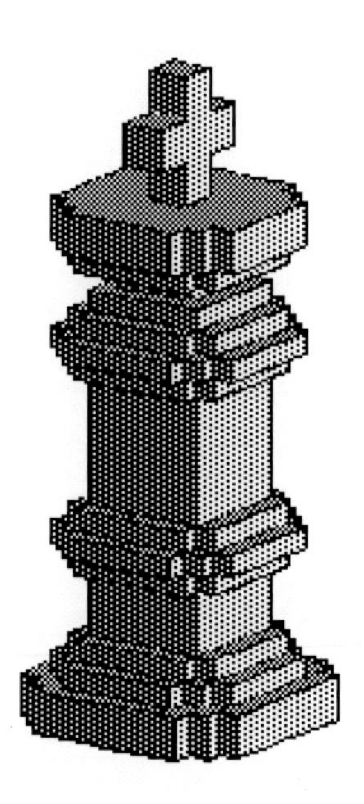

Fig. 2.10. An example of scene designed with QuickVox (M. Kergall)

Fig. 2.11. A solution for the property "Visibility is of 70 meters"

Fig. 2.12. Another solution for the property "Visibility is of 70 meters"

The modeler described in [30] uses fuzzy logic and allows generation of ambiences for existing landscape scenes. The modeler usually generates more than one solution. In Fig. 2.11 and Fig. 2.12 one can see examples of ambience modeling in a foggy landscape. These examples present two solutions of ambiance for the wished high level property "Visibility is of 70 meters".

J. Vergne [32] developed a declarative modeler allowing to generate wood textures from a high level description of the wished kind of wood, for example name and age of the tree. In Fig. 2.13 one can see two examples of wood textures generated by this modeler.

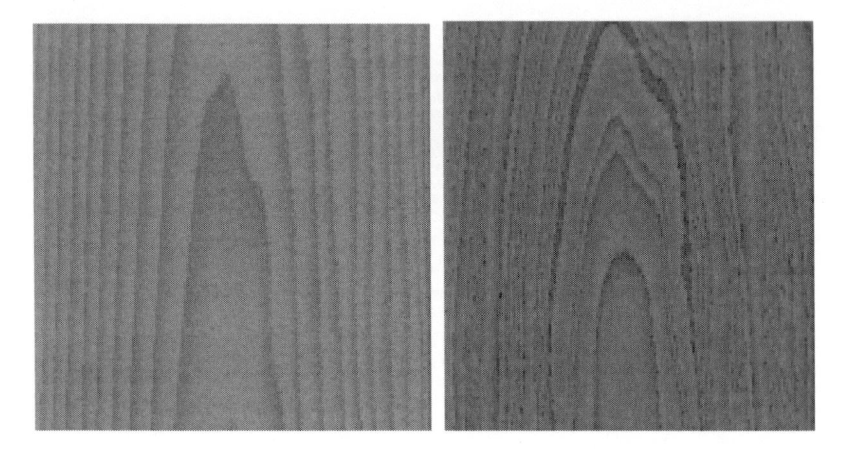

Fig. 2.13. Two wood textures generated by the modeler of J. Vergne

2.5 Scene Understanding in Declarative Scene Modeling

As declarative scene modeling generates several solutions and most of them can be unexpected, it is often necessary that the modeler offers scene understanding techniques in order to allow the designer to verify the properties of an obtained solution. Scene understanding can be visual or textual. Most of existing declarative modelers scene use simple scene display from an arbitrary chosen point of view. Very few declarative modelers use sophisticated scene understanding mechanisms.

PolyFormes uses a "matches-like" display mode allowing the user to better understand the shape of a generated polyhedron. In this kind of display, only the edges of the polyhedron are displayed but they are thickened (see Fig. 2.14).

Fig. 2.14. "Matches-like" display of polyhedrons

MultiFormes uses more sophisticated techniques for scene understanding. These techniques use a good view criterion based on the scene's geometry and automatically compute a good point of view by *heuristic search* [25]. Some other techniques for computing a good view have been proposed for particular cases [26].

As a single point of view is not always enough to understand complex scenes, MultiFormes also proposes an intelligent automatic scene exploration by a virtual camera,

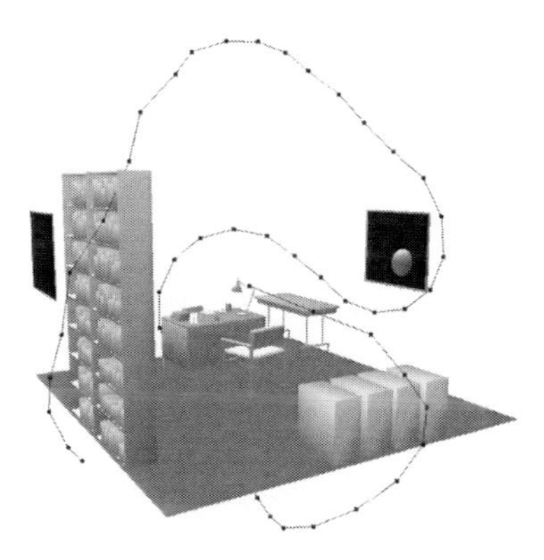

Fig. 2.15. Scene automated exploration by a virtual camera

moving on the surface of a sphere surrounding the scene [27, 28, 29, 39, 40]. An example of path of the virtual camera exploring a scene is shown in Fig. 2.15.

2.6 Constraint Satisfaction Techniques for Declarative Scene Modeling

In this section will be discussed one of the most popular techniques used in declarative scene modeling by the scene generation engine, in order to generate all the scenes verifying the properties wished by the user: resolution of a constraint satisfaction problem (CSP). To illustrate this technique, we will describe its implementation in a Declarative Modeling by Hierarchical Decomposition (DMHD) environment, more precisely in the Multiformes declarative scene modeler. Implementation of both arithmetic and geometric constraints will be discussed.

The high level properties of the scene given by the user during the description phase are translated to a set of constraints. In order to always get a finite number of solutions, numeric constraints upon discrete workspace are used.

2.6.1 Arithmetic Constraint Satisfaction Techniques

As each constraint has to be satisfied, the problem of the modeler is to find (at least) a set of values, one per variable, satisfying all the constraints. In other words, the modeler has to obtain solutions for a *constraint satisfaction problem* (CSP). In order to avoid some drawbacks of general form CSP and to obtain improvements due to the hierarchical decomposition, a new kind of primitive constraints and a new method to enforce problem reduction is used: constraint logic programming on finite domain, CLP(FD).

2.6.1.1 The Resolution Process

The resolution process for a CSP is an iterative process, each step being composed of two phases: the enumeration phase and the propagation phase.

1. The *enumeration phase* is made of two parts:

   - Selection of a variable;
   - Selection of a value for this variable.

   It is possible to use various orders in selecting variables and values. Some selection orders give best results than others. The variable selection order giving the best results it the order which follows the hierarchical decomposition tree given in the description phase.
2. The *propagation phase* takes into account the value of the variable chosen in the enumeration phase and tries to reduce the constraint satisfaction problem by propagating this value. Propagation of the value of the selected variable can be made in different manners:

   - Propagation through the affected constraints of the already instantiated variables. In this method, called "look-back", the resolution engine checks if the new value is compatible with the previously chosen values of variables.
   - Propagation through all variables, even the not yet instantiated. This kind of propagation is called "look-ahead" and can avoid future conflicts.

   If the "look-ahead" technique performs arc-consistency tests only between an instantiated and a not yet instantiated variable, it is called "forward-checking". If arc-consistency tests can be performed between not yet instantiated variables that have no direct connection with instantiated variables, in this case, the method is called "partial look-ahead".

2.6.1.2 Constraint Logic Programming on Finite Domains – CLP(FD)

The CLP(FD) method was initially proposed by Pascal Van Hentenryck [19, 20]. Several other authors contributed to its development.

The CLP(FD) method is based on the use of only primitive constraints of the form X **in** r, where X is a variable taking its values in a finite domain and r is a range. A range is a dynamic expression of the current domain of a variable and uses references to other variables via four *indexicals*, **min** (the min of the domain of the variable), **max** (the max of the domain of the variable), **dom** (the domain of the variable) and **val** (the value of the variable). Arithmetic expressions using indexicals, integer values and arithmetic operators can be used to describe a range.

In the final set of primitive constraints to be processed, each variable appears only once under the form: X **in** r1&r2&...&rn, where r1, r2, ..., rn are ranges and & denotes the union of ranges. This final form may be automatically obtained by applying rewriting rules. For instance, constraint "X=Y+Z" will give the set of primitive constraints:

X **in** min(Y)+min(Z) .. max(Y)+max(Z)
Y **in** min(X)-max(Z) .. max(X)-min(Z)
Z **in** min(X)-max(Y) .. max(X)-min(Y)

During the resolution process, a value is given to a variable X in order to reduce its range to a single value. The solver is then looking for indexicals using X in the domains of all the other variables and updates them. The ranges of the variables using these indexicals is reduced and this reduction is propagated to other variables. If the range of a variable becomes empty, there are no solutions for the chosen value of X and another value has to be tried.

With primitive constraints, various propagation strategies can be applied, depending on indexicals taken into account. Thus:

- Full look-ahead is obtained by using the indexical **dom**,
- Partial look-ahead is obtained by using only indexicals **min** and **max**. In this way, holes are not transmitted.
- Forward checking by using the indexical **val**, delaying the check of the constraint until the involved variable is ground.

2.6.1.3 Hierarchical Decomposition-Based Improvements

The CLP(FD) method can be more or less efficient, depending on the order to handle variables. The hierarchical decomposition tree used by DMHD gives the possibility to improve this order [14, 38].

Thus, the hierarchical decomposition tree can be processed by levels, the nodes of each level being rearranged in order to first visit simpler sub-trees, that is, sub-trees of less depth (Fig. 2.16). Exploration by levels and reordering permit to apply the "fail first" heuristic which tries to reach a possible failure as soon as possible.

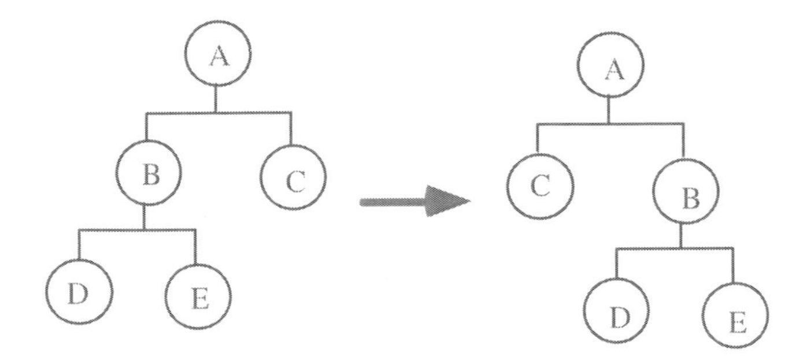

Fig. 2.16. Reordering according to the depth of subtrees

Another possibility to improve exploration by levels is to order the variables associated to a node of the hierarchical decomposition tree according to the number of constraints they share. This heuristic allows to process first, variables sharing a big number of constraints to reach sooner a possible failure.

A last possibility to improve resolution is to improve backtracking whenever a failure is reached. Let us consider the case of the tree of Fig. 2.17. During the processing of the level of nodes D, E, F and G, when a failure occurs in nodes E, F or G, backtracking is classically performed to the previous node (D, E or F respectively).

However, values of variables of nodes D and E are completely independent from values of variables of nodes F and G. If a failure occurs in node F or G, nodes D and E are not responsible for this failure. With intelligent backtracking, if a failure occurs in nodes F or G, backtracking is performed to the last node of the previous level. So,

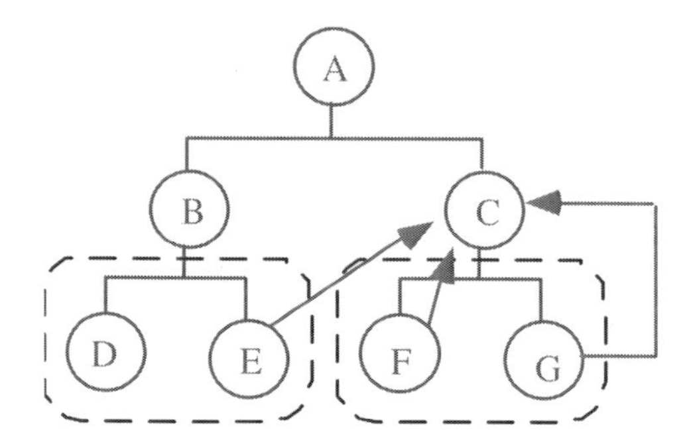

Fig. 2.17. Grouping of nodes and intelligent backtracking

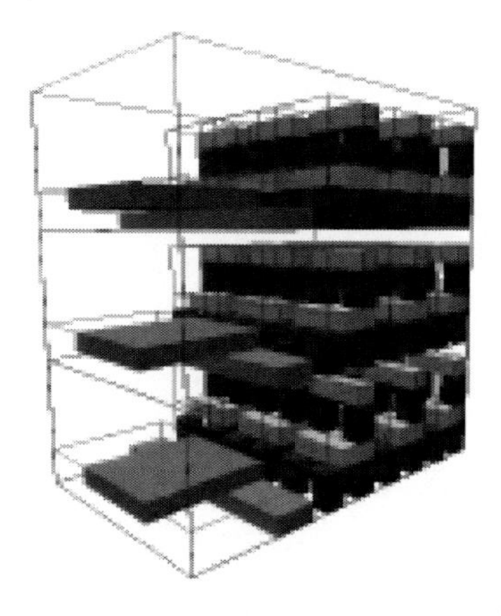

Fig. 2.18. Inside a 3 floor building

for the example of Fig. 2.17, backtracking is performed to node C, without exploring all the remaining possibilities for nodes D and E.

Intelligent backtracking is based on independence of various parts of the CSP and these parts are easy to determine with the hierarchical decomposition tree. This independence has been exploited to parallelize the resolution process in MultiFormes. Fig. 2.18 shows an example of a scene generated with the arithmetic CSP solver of MultiFormes.

2.6.2 Geometric Constraint Satisfaction Techniques

The main drawback of the use of mainly linear arithmetic constraints to describe properties in MultiFormes is that decomposition is always performed by using bounding boxes of the various parts of the scene, the faces of these boxes being parallel to the plans defined by the main axes. This constraint limits the expressiveness of the modeler, especially in reusing of already existing scenes.

In order to improve the expressiveness of MultiFormes, a solver of geometric CSP has been defined and developed.

2.6.2.1 Principles of the MultiFormes Geometric Constraint Solver

The geometric constraint solver of MultiFormes is based on CSP solving techniques. The main difference with classical CSP solvers is that now, variables are replaced by points. A point is the smallest entity handled by the solver.

More precisely, a *geometric CSP* will be defined as a triplet (V, D, C) where V is a set of variable points (or vertices), D is a set of 3D sub spaces, domains of the points of the set V, and C is a set of geometric constraints on points. A solution of a geometric CSP is a set of positions in the 3D space verifying all the constraints and belonging to the domain of each variable point.

For reasons of compatibility with the arithmetic constraint solver of MultiFormes, constraints of a geometric CSP are binary constraints of the form X **in** S, where X is a variable point and S a restriction of the 3D space.

All the high level properties used by the modeler have to be translated to a set of binary geometric constraints. For instance, the property "The three points A, B and C are aligned" will be translated to the set of binary geometric constraints:

A **in** (BC),
B **in** (AC),
C **in** (AB).

2.6.2.2 The Resolution Process

The resolution process is based on a classical chronological backtrack algorithm. Variable points are sequentially processed. The algorithm tries to find a value in the domain of the current variable point, satisfying the constraints with the current values of already processed variable points. In case of failure, another value of the current variable point is reached. In case of failure for any possible value of the current variable point, backtracking is performed to the last processed variable point and a new value is tried for this point.

The domain of a variable point is computed by computing the intersection of all the domains assigned to the variable point by properties including this point. If the final domain is empty, there is no solution for this variable point, that is there no position for this point satisfying all its constraints. If the obtained final domain is continuous, it has to be sampled.

2.6.2.3 The Intersection and Sampling Problems

The resolution process uses intersection computation of geometric objects as spheres, boxes, straight lines etc. The intersection computation process may produce either a discrete or a continuous domain. When a discrete domain is obtained, all the possible discrete values have to be tried by the solution searching process. Otherwise, the obtained continuous domain must be sampled in order to get a finite number of solutions.

The geometric entities used by the solver are: point, straight line, half-straight line, segment, plan, convex polygon, box, sphere, circle, arc of circle and ball (solid sphere). However, the intersection of some geometric entities may produce an entity not present in the above list. In such a case approximation is used in order to get a known kind of geometric entity. The problem is that the sampling process, applied to an approximated domain, may produce points which are not solutions because they don't belong to the real domain. Thus, a filtering step has been added in the resolution

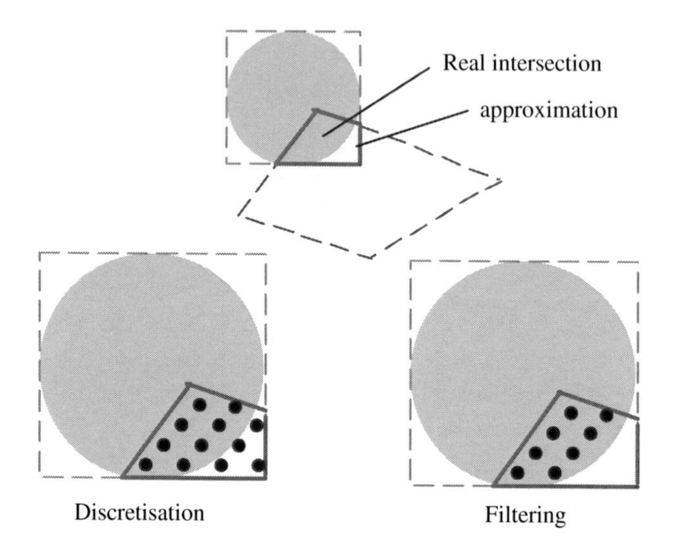

Fig. 2.19. Approximation, sampling and filtering of an intersection

process, in order to verify that a chosen value belongs to the real domain. In Fig. 2.19, one can see illustration of this problem for the case of intersection of a circle and a convex polygon.

Approximation and sampling of intersections have to be delayed as late as possible in order to avoid problems of lack of solutions which may occur if an approximated discrete domain is intersected by another approximated discrete domain. A special mechanism of the geometric CSP solver rearranges intersection computations in order to use approximation and sampling only at the end of the intersection computation process.

2.6.2.4 Some Other Problems

The following two problems may affect the efficiency of the geometric CSP solver.

The first one is due to the need of sampling. Existence of solutions may depend on the instantiation order of the variable points. For example, let us consider the case where two points C and D have to belong to the already instantiated segment (AB) and BD has to be 4 times longer than AC. If the sampling mechanism gives three possible positions on the segment (AB), the following cases may occur:

- The variable point C is instantiated before the point D. None of the three possible positions of C is a solution because, the constraint BD = 4 * AC is not satisfied for any position of D (Fig. 2.20, left).
- The variable point D is instantiated before the point C. It is possible to find a solution for this problem (Fig. 2.20, right).

Another problem is the one of missing constraints. For the user, transitivity is an obvious property. So, if he (she) has to tell the modeler that segments (AB), (CD) and (EF) must have the same length, he (she) may think that it is enough to tell that (AB) = (EF) and (CD) = (EF). In fact, a constraint is missing, because the modeler does not

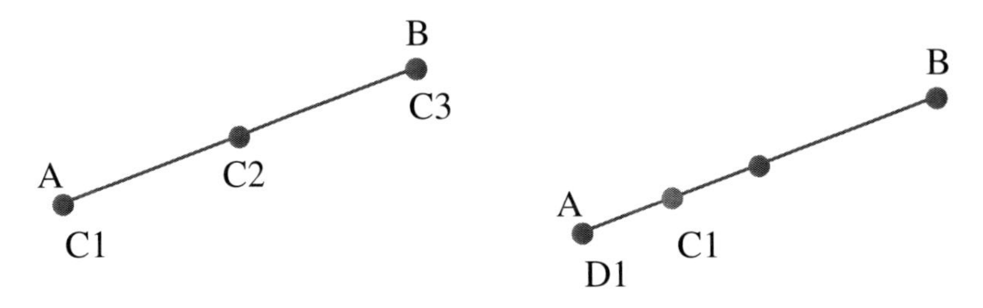

Fig. 2.20. The instantiation order problem. If C is instantiated before D, the solver doesn't find solution, even if a solution exists.

Fig. 2.21. On the left: a table and 4 chairs; on the right: The collegiate church of Le Dorat (France)

know transitivity. In order to complete the description, a term rewriting system based on the Knuth-Bendix algorithm has been added to the solver.

Fig. 2.21 shows two scenes results obtained by the geometric CSP solver of the MultiFormes declarative scene modeler. The first scene represents a table and 4 chairs and the second one, the collegiate church of Le Dorat, in France. In both scenes, one can see reusing of scene elements described only once.

2.6.3 Discussion

The use of CSP resolution techniques in declarative modeling,, together with hierarchical decomposition, improves the efficiency and expressivity of declarative modelers. However, several drawbacks remain when using either arithmetic or geometric CSP. A brief recall on advantages and drawbacks of arithmetic and geometric CSP is made in this section.

2.6.3.1 Arithmetic CSP

The main advantage of the use of special arithmetic CSP on finite domains is that, combined with hierarchical decomposition in declarative modeling, it is an efficient tool allowing reduction of workspace and solution space, in order to obtain solutions very quickly.

However, the use of arithmetic CSP on finite domains has several drawbacks. The most important of these drawbacks are the following:

- Due to the use of sampling, some solutions may be lost.
- As the solver processes mainly linear constraints, it is difficult to handle easily objects with sides non parallel to the 3 main plans.

2.6.3.2 Geometric CSP

The main advantage of the use of geometric CSP in declarative modeling is that this kind of CSP allow to process various objects and not only boxes with sides parallel to the main plans. Thus, the expressiveness of the declarative modeler is increased.

As it has been written in section 4, some drawbacks are specific to geometric CSP:

- Even if sampling is performed as late as possible, it may conduct to situations where no solution is found, whereas there exist solutions.
- Existence of solutions may depend on the instantiation order of variable points.

Maybe, the solution is in efficient cooperation between a declarative modeler, combining both arithmetic and geometric constraints and a classical geometric scene modeler. In such a case, the declarative modeler would allow to create sketches of scenes which would be refined by the classical geometric modeler. Communication between the two modelers could be bidirectional, allowing to submit to the declarative process a scene obtained with the classical geometric modeler, in order to be generalized [17, 18, 43].

2.7 Declarative Scene Modeling and Machine-Learning Techniques

We have seen that a declarative scene modeler can be used in two different modes: *exploration* mode and *solution search* mode.

In exploration mode, due to the use of imprecise properties, the modeler may generate a very important number of solutions. Even if the generation of all the solutions increases the richness of the modeler, it may be a serious drawback because the modeler may generate solutions which are of no interest for the user. In order to avoid this kind of proposed non interesting solutions machine-learning can be proposed in order to learn the modeler the user's preferences, that is, what kind of scenes are, or are not, interesting.

In solution search mode, the designer has a relatively precise idea of the kind of scenes he (she) would like to obtain. Thus, the designer would like to obtain a solution immediately or very quickly from a description using less imprecise properties. Unfortunately the semantic of a property is often ambiguous and several solutions not satisfactory for the user can be faced by the modeler. In such cases, the knowledge of the designer's preferences can guide the modeler in its search. So, if some proposed solutions do not satisfy the designer, it would be interesting to learn the modeler not to examine this kind of solutions. This learning decreases the solution space because, for a great number of scenes verifying the properties of the initial description, some scenes will not satisfy the intuitive idea of the user and these scenes will be avoided.

It is easy to see that machine learning often lightens the designer's work and increases the efficiency of generation. In the following subsections we present two methods to implement machine learning for declarative modeling by hierarchical decomposition in order to improve modeling in both exploration and solution search mode.

2.7.1 A Dynamical Neural Network for Filtering Unsatisfactory Solutions in DMHD

In order to improve exploration mode generation, but also solution mode generation,an interactive machine learning mechanism has been implemented, based on the use of neural networks [8, 9, 10, 11, 13], applied to DMHD. This mechanism is used by the modeler in the following manner:

- During a learning phase, some scenes, generated by the modeler from the initial description, are selected by the user to serve as examples of desired scenes. Each time a new example is presented, the modeler learns more on the user's preferences and this is materialized by a modification of the values of weights associated to the connections of the network. At the end of the learning phase, an acceptation interval is calculated and assigned to each decision cell.
- After the end of the learning phase, the modeler is in the normal working phase: the weights of connections calculated during the learning phase are used as filters allowing the choice of scenes which will be presented to the user.

The used machine learning mechanism takes into account only relative dimension and position properties of a scene. Form properties are processed by the scene generation engine after selection of dimensions and positions of the bounding boxes of each sub-scene.

2.7.1.1 Structure of the Used Network

The neural network is created dynamically from the description of the scene and its structure is described in the following lines.

To each node of the scene description tree, are assigned:

- A network whose the input layer is composed of two groups of neurons (see figure 2.22) : a group of two neurons whose inputs are w/h and w/d where w, h and d are respectively the width, the height and the depth of the scene associated with the current node; a group of neurons whose inputs are the results of acceptance of the other nodes of the scene. The network contains another layer of two neurons which work in the following manner: the first one evaluates the quantity i1*w1+ i2*w2, where wk represents the weight attributed to the connection between the k-th neuron of the input layer and the intermediate layer and ik the k-th input; it returns 0 or 1 according to whether this weighted sum belongs to a given interval or not. The values of weights wk can be modified during the learning phase. The second neuron computes the sum of acceptance results coming from the second group of neurons of the input layer. Its output function returns 0 or 1 according to whether this sum is equal or not to the number of acceptance neurons. The decision layer of the network contains one neuron and computes the sum of the outputs of neurons, returning 1 (acceptance) or 0 (refusal) according to whether this sum is equal to 2 or not.

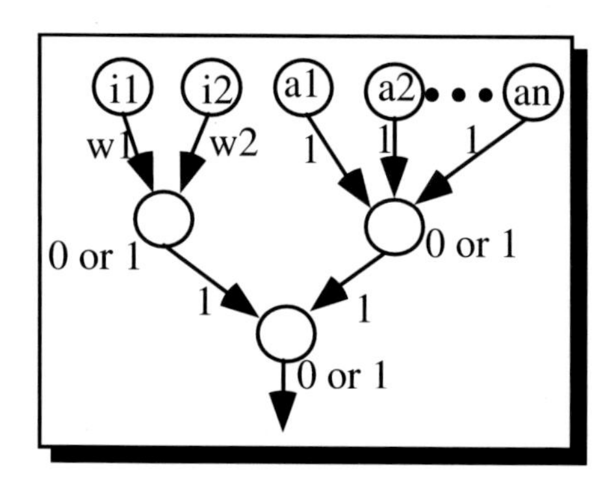

Fig. 2.22. Local neural network

- n*(n-1)/2 networks, corresponding to all possible arrangements of two sub-scenes of the current scene, where n is the number of child-nodes of the current node. These networks have the same structure and the same behavior as the others, excepting that the inputs of the first group of neurons of the input layer are dx/dy and dx/dz, where dx, dy and dz are the components of the distance d between the two sub-scenes. The values of weights wk can be modified during the learning process.

Let us consider the scene House, described by the following Prolog-like pseudo-code:

```
House(x) ->
       Habitation(x1)
       Garage(x2)
       PasteLeft(x1,x2);

Habitation(x) ->
       HigherThanWide(x)
       HigherThanDeep(x)
       Roof(x1)
       Walls(x2)
       PutOn(x1,x2);

Garage(x) ->
       TopRounded(x,70);

Roof(x) ->
       Rounded(x,60);

Walls(x,p) ->;
```

The generated neural network for this description is shown in figure 2.23.
The acceptance process of a scene generated by the modeler is the following:

- If the current node of the generated scene is accepted at the current level of detail and relative placements of all the sub-scenes of the node at the next level of detail are also accepted by the associated neural networks, the scene is partially accepted at the current level of detail and acceptance test is performed with each child-node of the current node. A scene is accepted at the current level of detail if it is partially accepted for each node up to this level of detail.
- A scene generated by the modeler is accepted if it is accepted at each level of detail. Otherwise, the scene is refused and it is not presented to the user.

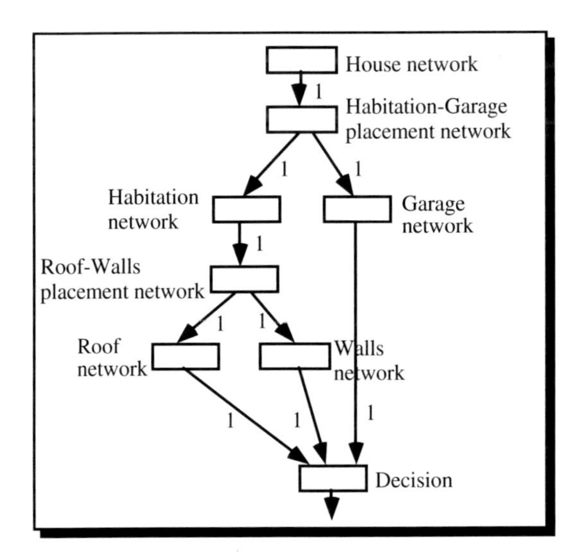

Fig. 2.23. Neural network generated for the "House" scene description

2.7.1.2 The Machine Learning Process

For each neural network of the general network, machine learning is performed in the same manner.

From each new example of scene presented by the user among the scenes generated by the modeler, each local network learns more and more and this fact is materialized by modifying the weights of connections between the association (intermediate) layer and the decision (output) layer. The formula used to adjust a weight w is the following:

$$wi+1 = w0 - wi\ V(Xi+1)$$

where
wi represents the value of the weight w at step i,
w0 represents the initial value of the weight w,
Xi represents the value of the input X at step i,
V(Xi) represents the variance of values X1, ..., Xi.

This formula permits to increase, or not to decrease, the value of the weight when the corresponding input value is constant and to decrease the value of the weight when the corresponding input value varies.

When, at the end of the learning phase, the final value of the weight w has been computed, the value of the quantity (w - w0) / m is estimated, where m is the number of examples selected by the user. If the value of this quantity is less than a threshold value, the weight w takes the value 0; otherwise, it takes the value 1.

The quantity computed by the output function of the first neuron of the intermediate layer of each local network is:

$$S = w1\ Xi + w2\ Yi,$$

where Xi and Yi are the input values at step i.

Thus, at the end of the machine learning phase, each output function of the first neuron of the intermediate layer of each local network will compute the quantity:

$$Sa = w1\ A(X) + w2\ A(Y)$$

where A(X) represents the average value of the input values X1, ..., Xn.

During the phase of normal working of the modeler, a solution is partially accepted by a local neural network if the output value computed by the output function of the first neuron of the intermediate layer belongs to the neighboring of the value Sa.

Let us consider three cases for each local network:

1. w1 = w2 = 0.

The value of Sa is then equal to 0 and all solutions are accepted.

2. w1 = 0, w2 = 1.

The value of Sa is then equal to A(Y). Only the input value associated with w2 is important and only scenes which have, for their corresponding to the current local network part, input values close to the value of A(Y) are selected.

3. w1 = w2 = 1.

The value of Sa is then equal to A(X) + A(Y). The two quantities associated with w1 and w2, are important. The parts of scenes selected by the current local neural network will be the ones whose input values associated with w1 are close to A(X) and input values associated with w2 are close to A(Y), but also other scenes for which the sum of entry values is equal to A(X) + A(Y) whereas they should not be accepted.

2.7.1.3 Discussion

Although neural networks are fully efficient for linearly separable problems, their application to declarative modeling for a non linearly separable problem, selecting scenes close to those wished by the designer, gives results globally satisfactory for exploration mode generation, because it lightens the designer's work by filtering the major part of non interesting solutions. Machine learning is performed with little information and already learnt knowledge can be used for continuous machine learning where the modeler avoids more and more non interesting scenes.

Figure 2.24 shows some rejected scenes during the machine learning phase for the above "House" scene description. The purpose of this phase was to learn that the "Habitation" part of the scene must be wider than the garage part. In figure 2.25, one

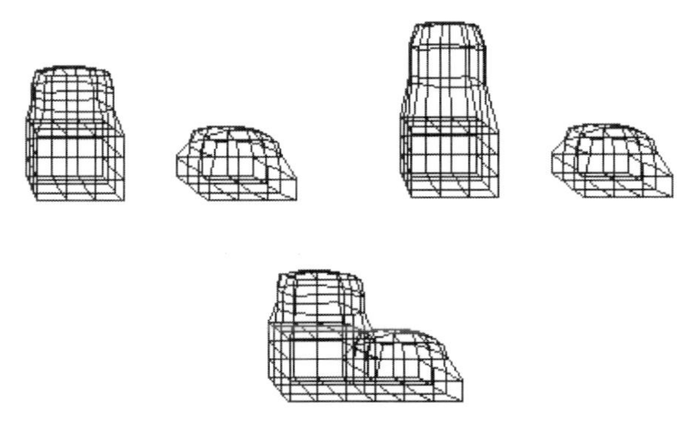

Fig. 2.24. Scenes generated from the "House" description and rejected by the user

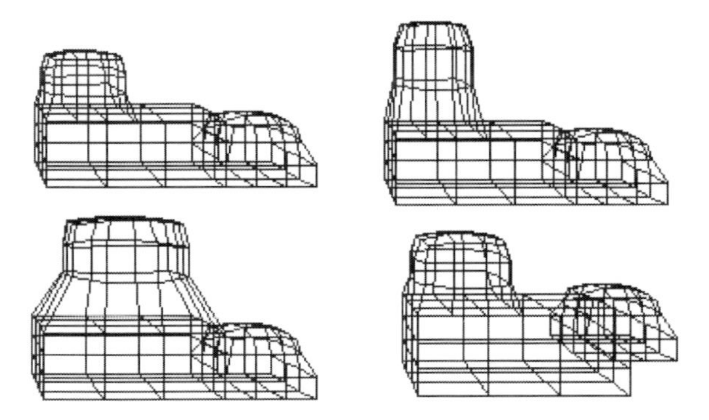

Fig. 2.25. Machine learning for the "House" description. Only generated scenes with the wished shape are presented to the user.

can see some generated scenes for the "House" scene description, after the machine learning phase.

Of course, this kind of interactive machine learning reduces the number of solutions shown to the user but not the number of tries; the total exploration time remains unchanged. Another drawback of the method is that learning is not efficient if the number of given examples is small and that learning can be avoided if the given examples are not well chosen.

We think that it is possible to use the same neural network to reduce the number of tries during generation. To do this, after the machine learning phase, the neural network has to be activated as soon as possible, after each enumeration phase in the constraint satisfaction process. In this manner, the next step of the resolution process will take place only if the scene goes through the neural network filter.

Some other machine-learning techniques were proposed [36, 23] based on the use of genetic algorithms [37] or on other classification techniques.

2.8 Advantages and Drawbacks of Declarative Scene Modeling

The purpose of declarative scene modeling is to improve the hard task of scene modeling by allowing the designer to use a high level of abstraction. Taking into account the existing declarative modelers, it is possible to extract the main advantages and drawbacks of declarative modeling.

Declarative modeling is closer to the user than traditional geometric modeling because the user has not to take into account geometric representation and construction of the scene. The declarative modeler produces a scene with the high level properties described by the designer and then translates this solution to a geometric model.

In declarative modeling it is possible to describe a scene up to a chosen level of detail. In this manner, even if the designer has not yet determined some parts of the scene, he (she) can get a first draft of the scene and refine it later.

Modifications of a scene are easier with declarative modeling because it is easier to replace a property by another one than to modify the scene's geometry. Indeed, it is easier to modify an object by telling the modeler that it is not enough high than by changing the list of vertices in the geometric model of the object.

Declarative modeling allows really computer aided design because, from an initial description of a vague idea of the designer it is able to propose several solutions and help the designer in the creative phase of scene design.

A declarative modeler can evolve because, if properties are defined in a manner not depending on the scene generation engine, new properties can easily be added to the modeler and improve its designing power.

The main drawback of declarative modeling comes from the possibility to use imprecise properties. When, for a given description, the *search space* is much wider than the *solution space*, the generation process is very time consuming. This is especially true with general purpose declarative modelers. A research field in the area of declarative modeling is to find efficient methods of reducing the search space.

Another drawback is due to the possibility to get many solutions from a scene description. If this possibility is an advantage of declarative modeling, the management of an important number of solutions is always difficult because the designer cannot remember all the solutions during the designing process. A possible solution would be to define a metric and classes of close to each other solutions by the modeler and to show only a representative scene from each class. The problem is that it is not always possible to define such a metric, especially with general purpose modelers.

2.9 Future Issues

The problem declarative modeling has to face is a hard one. Despite of this complexity, the performances of existing declarative modeler prototypes are quite satisfactory. However, the problem of efficient reduction of the search space is an open research problem.

Current declarative modelers are essentially concerned with geometrical or topological aspects of a scene. However, it is possible to describe in declarative manner non geometric properties of the scene such as ambience (lighting, fog, ...). Some research works have started in this area. In a general manner, if a property may be translated into constraints on numerical values, it can be described and processed in declarative manner.

Another challenge is to include time in declarative modeling, especially for scenes corresponding to existing things in the real world. For instance, it would be interesting for an architect to show a client not only the current state of a building to be designed but also its appearance 10 or 15 years later, if a small number of parameters such as, construction materials, climate, etc., are known. We call this kind of modeling *predictive declarative modeling*. Some work has started in this area too.

Finally, coupling declarative and traditional scene modeling should give interesting results because it would allow to reduce the cost of declarative modeling, by permitting to first define a draft of the scene by declarative modeling and then refine the draft using an integrated geometric modeler. Such integrated declarative-traditional geometric modelers have already been implemented [17, 18, 43] but a lot of improvements are still possible in this area.

2.10 Conclusion

In this chapter we have tried to present the challenge of declarative modeling as a tool to make easier the scene designer's work and offering a really computer aided design able to stimulate the designer's creativity. Even if many things have to be improved in declarative modeling, most dedicated declarative modelers are very efficient and, above all, able to produce scenes impossible to obtain by traditional geometric modeling.

Declarative modeling has opened a lot of research fields in computer graphics as well as in artificial intelligence. Open research fields currently concern reduction of search space during the generation phase (efficient constraint resolution), management of a big number of solutions (classification, machine-learning, interactivity) and scene understanding (scene visual complexity, path optimization).

References

1. Lucas, M., Martin, D., Martin, P., Plemenos, D.: The ExploFormes project: some steps towards declarative modeling of forms. In: AFCET-GROPLAN Conference, Strasbourg, France, November 29 - December 1, 1989, vol. (67), pp. 35–49. BIGRE (1989) (in French)
2. Plemenos, D.: A contribution to study and development of scene modeling, generation and display techniques - The MultiFormes project. Professorial Dissertation, Nantes, France (November 1991) (in French)
3. Plemenos, D.: Declarative modeling by hierarchical decomposition. The actual state of the MultiFormes project. In: International Conference GraphiCon 1995, St Petersbourg, Russia, July 3-7 (1995)
4. Plemenos, D., Tamine, K.: Increasing the efficiency of declarative modeling. Constraint evaluation for the hierarchical decomposition approach. In: International Conference WSCG 1997, Plzen, Czech Republic (February 1997)
5. Lucas, M., Desmontils, E.: Declarative modelers. Revue Internationale de CFAO et d'Infographie 10(6), 559–585 (1995) (in French)
6. Martin, D., Martin, P.: An expert system for polyhedra modeling. In: EUROGRAPHICS 1988, Nice, France (1988)
7. Martin, D., Martin, P.: PolyFormes: software for the declarative modeling of polyhedra. The Visual Computer, 55–76 (1999)
8. Mc Culloch, W.S., Pitts, W.: A logical calculus of the ideas immanent in nervous activity. Bulletin of Mathematical Biophysics 5, 115–133 (1943)

9. Rosenblatt, F.: The perceptron: a perceiving and recognizing automaton. Project Para, Cornell Aeronautical Lab. Report 85-460-1 (1957)
10. Plemenos, D.: Techniques for implementing learning mechanisms in a hierarchical declarative modeler. Research report MSI 93 - 04, Limoges, France, (December 1993) (in French)
11. Boughanem M., Plemenos D.: Learning techniques in declarative modeling by hierarchical decomposition. In: 3IA 1994 Conference, Limoges, France, April 6-7 (1994) (in French)
12. Plemenos, D., Ruchaud, W., Tamine, K.: New optimising techniques for declarative modeling by hierarchical decomposition. In: AFIG annual conference, December 3-5 (1997) (in French)
13. Plemenos, D., Ruchaud, W.: Interactive techniques for declarative modeling. In: International Conference 3IA 1998, Limoges, France, April 28-29 (1998)
14. Bonnefoi, P.-F.: Constraint satisfaction techniques for declarative modeling. Application to concurrent generation of scenes. PhD thesis, Limoges, France (June 1999) (in French)
15. Kochhar, S.: Cooperative Computer-Aided Design: a paradigm for automating the design and modeling of graphical objects. PhD thesis, Harvard University, Aiken Computation Laboratory, 33 Oxford Street, Cambridge, Mass. 02138, available as TR-18-90 (1990)
16. Kochhar, S.: CCAD: A paradigm for human-computer cooperation in design. IEEE Computer Graphics and Applications (May 1994)
17. Sellinger D., Plemenos D.: Integrated geometric and declarative modeling using cooperative computer-aided design. In: 3IA 1996 conference, Limoges, France, April 3-4 (1996)
18. Sellinger, D., Plemenos, D.: Interactive Generative Geometric Modeling by Geometric to Declarative Representation Conversion. In: WSCG 1997 conference, Plzen, Czech Republic, February 10-14 (1997)
19. van Hentenryck, P.: Constraint satisfaction in logic programming. Logic Programming Series. MIT Press, Cambridge (1989)
20. Diaz, D.: A study of compiling techniques for logic languages for programming by constraints on finite domains: the clp(FD) system
21. Chauvat, D.: The VoluFormes Project: An example of declarative modeling with spatial control. PhD Thesis, Nantes (December 1994) (in French)
22. Poulet, F., Lucas, M.: Modeling megalithic sites. In: Eurographics 1996, Poitiers, France, pp. 279–288 (1996)
23. Champciaux, L.: Introduction of learning techniques in declarative modeling, PhD thesis, Nantes, France (June 1998)
24. Kwaiter, G.: Declarative scene modeling: study and implementation of constraint solvers, PhD thesis, Toulouse, France (December 1998)
25. Plemenos, D., Benayada, M.: Intelligent display in scene modeling. New techniques to automatically compute good views, GraphiCon 1996, St Petersburg, Russia, July 1-5 (1996)
26. Colin, C.: Automatic computing of good views of a scene. In: MICAD 1990, Paris, France (February 1990)
27. Barral, P., Dorme, G., Plemenos, D.: Visual understanding of a scene by automatic movement of a camera, GraphiCon 1999, Moscow, Russia, August 26 - September 1 (1999)
28. Barral, P., Dorme, G., Plemenos, D.: Visual understanding of a scene by automatic movement of a camera. In: Eurographics 2000 (2000)
29. Barral, P., Dorme, G., Plemenos, D.: Visual understanding of a scene by automatic movement of a camera. In: International conference 3IA 2000, Limoges, France, March 3-4 (2000)

30. Jolivet, V., Plemenos, D., Poulingeas, P.: Declarative specification of ambiance in VRML landscapes. In: ICCS 2004 international conference (CGGM 2004), Krakow, Poland, June 6-9, 2004. LNCS, pp. IV115–IV122. Springer, Heidelberg (2004)
31. Fribault, P.: Declarative modeling of livable spaces. Ph. D thesis, Limoges, France, November 13 (2003) (in French)
32. Vergne, J.: Declarative modeling of wood textures. MS thesis, Saint-Etienne, France (June 2003) (in French)
33. Kergall, M.: Scene modeling by three shadows. MS thesis, Rennes, France (September 1989) (in French)
34. Bonnefoi, P.-F., Plemenos, D.: Constraint satisfaction techniques for declarative scene modeling by hierarchical decomposition. In: 3IA 2000 International Conference, Limoges, France, May 3-4 (2002)
35. Ruchaud, W.: Study and realisation of a geometric constraint solver for declarative modeling. PhD Thesis, Limoges, France, November 15 (2001) (in French)
36. Vassilas, N., Miaoulis, G., Chronopoulos, D., Konstantinidis, E., Ravani, I., Makris, D., Plemenos, D.: MultiCAD GA: a system for the design of 3D forms based on genetic algorithms and human evaluation. In: SETN 2002 Conference, Thessaloniki, Greece (April 2002)
37. Goldberg, D.E.: Genetic Algorithms. Addison-Wesley, Reading (1991)
38. Bonnefoi, P.F., Plemenos, D., Ruchaud, W.: Declarative modeling in computer graphics: current results and future issues. In: ICCS 2004 international conference (CGGM 2004), Krakow, Poland, June 6-9, 2004. LNCS, pp. IV80-IV89. Springer, Heidelberg (2004)
39. Sokolov, D., Plemenos, D., Tamine, K.: Methods and data structures for virtual world exploration. The Visual Computer 22(7), 506–516 (2006)
40. Plemenos, D., Sbert, M., Feixas, M.: On viewpoint complexity of 3D scenes. STAR Report. In: International Conference GraphiCon 2004, Moscow, Russia, September 6-10, 2004, pp. 24–31 (2004)
41. Bardis, G.: Machine-learning and aid to decision in declarative scene modeling. Ph. D thesis, Limoges (France), June 26 (2006)
42. Makris, D.: Study and realisation of a declarative system for modeling and generation of styles by genetic algorithms. Application to architectural design. Ph. D thesis, Limoges (France), October 18 (2005)
43. Golfinopoulos, V.: Study and realisation of a knowledge-based system of reverse engineering for declarative scene modeling. Ph. D thesis, Limoges, France, June 21 (2006)
44. Dragonas, J.: Collaborative declarative modeling. Ph. D thesis, Limoges, France, June 26 (2006)
45. Miaoulis, G.: Study and realisation of an intelligent multimedia information system for computer-aided design. The MultiCAD project. Ph. D thesis. Limoges, France (February 2002)

3 Understanding Scenes

Vassilios S. Golfinopoulos

Department of Computer Science, Technological Education Institute of Athens,
Ag.Spyridonos St., 122 10 Egaleo, Greece
golfinopoulos@teiath.gr

Abstract. Declarative modelling allows the designer to describe a scene, without the need to define the geometric properties, by specifying its properties which can be imprecise and incomplete. Declarative modelling by hierarchical decomposition is a special approach which gives the user the ability to describe a scene by top-down decomposition at different levels of detail, generates a set of geometric solutions that meet the description and visualizes the scenes. The aim of the present work is to settle the reverse engineering process and by exploiting knowledge couples a declarative with a traditional geometric modeller. The declarative conception cycle of declarative modelling is extended, in order to include the reverse engineering process, by introducing the reconstruction phase and the iterative design process becomes automated. The reconstruction phase receives a set of selected scenes, which are semantically understood, permits the designer to perform geometric and topological modifications on the scene(s) and results an abstract description which embodies the designer modifications and leads to more promising solutions, reducing the initial solution space.

Keywords: Declarative modelling, Reverse engineering, Knowledge-based systems, Computer-aided design.

3.1 Introduction to Reverse Engineering

Engineering is the process involved in designing, manufacturing, constructing, and maintaining of products, systems, and structures. At a higher level, there are two types of engineering: forward engineering and reverse engineering.

Generally speaking reverse engineering is the process of taking something (a device, an electrical component, a software program, et cetera) apart and analyzing its workings in detail, usually with the intention to construct a new device or program that does the same thing without actually copying anything from the original. In other words, reverse engineering is the process of analyzing a subject system to (i) identify the system's components and their interrelationships and (ii) create representations of the system in another form or a higher level of abstraction [5].

Forward engineering is the traditional process of moving from high-level abstractions and logical, implementation-independent designs to the physical implementation of a system [5]. Reengineering (also known as renovation and reclamation) is the examination and alteration of a subject system to reconstitute it in a new form and the subsequent implementation of the new form. Reengineering is the modification of a

G. Miaoulis and D. Plemenos (Eds.): Intel. Scene Mod. Information Systems, SCI 181, pp. 59–87.
springerlink.com

system that takes place after it has been reverse engineered, generally to add new functionality.

Reverse engineering is very common in such diverse fields as software engineering, entertainment, automotive, consumer products, microchips, chemicals, electronics, and mechanical designs. In some situations, designers give a shape to their ideas by using clay, plaster, wood, or foam rubber, but a CAD model is needed to enable the manufacturing of the part. As products become more organic in shape, designing in CAD may be challenging or impossible. There is no guarantee that the CAD model will be acceptably close to the sculpted model. Reverse engineering provides a solution to this problem because the physical model is the source of information for the CAD model.

3.1.1 Reverse Engineering in Scene Modelling

As mentioned before, the design process can be viewed as a successive transformation between models. These models can be classified into a representation hierarchy of most general to most specific according to the levels of abstraction. Usually, we start the design process indicating a conceptual model which is transformed into specific model. Within the framework of MultiCAD, several scene model types exist according to levels of abstraction [21]:

- Abstract/conceptual models: The abstract models are scene models, necessarily generic and can give a series models of lower level of abstraction. The type of these models can be internal declarative representations or documents in natural language or models based on semantic networks.
- Intermediate models: They are primarily geometric models. These models are concrete from the point of view of the geometric properties of objects, of their composition and of their of space arrangement. Geometrical models can produce several different models of the lower level of abstraction if for example differ the properties of appearance such as the color or texture of the objects.
- Physical models: The models of the physical level contain information appropriate to the direct visualization of the 3D objects that constitute the scene. They contain all necessary object properties in order to represent real situations.

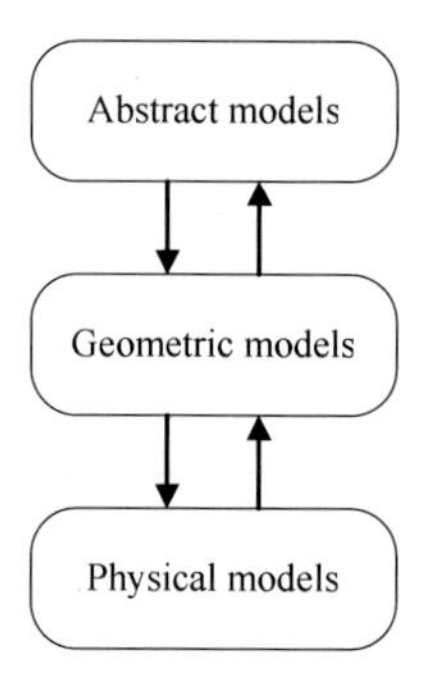

Fig. 3.1. The transformation of models

[21] introduces the notion of vertical transformation of the levels of abstraction. An abstract model is transformed or generates a set of concrete models in terms of more concretion and less ambiguity. The reverse process of the generation is the comprehension where a physical model is transformed into geometric and abstract model with less concretion and more ambiguity. Figure 3.1 illustrates the transformation of the models. The comprehension process is actually the reverse engineering within the framework of the design process.

Reverse engineering is an important branch of the design process and has been widely recognized as a crucial step in the product design cycle. Differing from the traditional design idea and method, reverse engineering technology enables one to start a design from an existing product model by combining computer technology, measurement technology and CAD/CAM technology [34]. In the forward design process the operation sequence usually starts from an idea, an abstract model via computer-aided design (CAD) techniques, and ends with generation of the geometric model that represents the initial idea. In contrast to this conventional design process, reverse engineering represents an approach for the new design of a product that may lack an existing CAD model. In the process of the product design and research, the use of reverse engineering will reduce the production period and costs. Reverse engineering technology is not to copy an existing product but to acquire a design concept from an existing physical model and create a complete geometric model and further to optimize the product design.

According to [23] the application areas of a reverse engineering process include:

- The reverse design: either creating a new product from an initial model or feeding a recovered result back to an existing product model to compare and update. This is a widely used technique in the tool, die, and mould-making industries.
- The customized design: customized products are worn on our bodies, or have prolonged functional contact with the human body. There can be considerable variation in performance and function required for this kind of products and it is, therefore, essential to involve the customer in the design process.
- The virtual environment: to build the virtual reality environment in which the overall design of a product can be evaluated quickly and effectively.

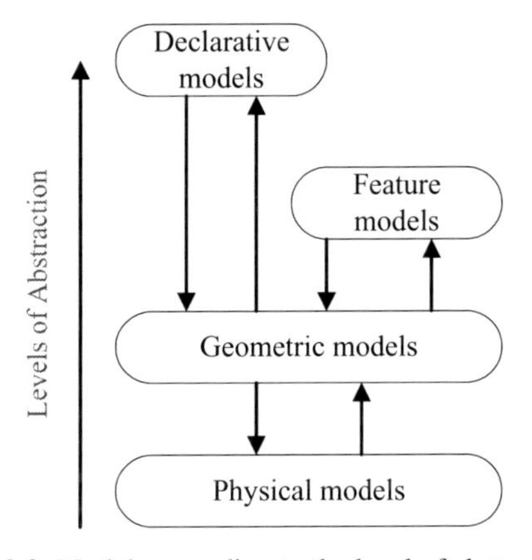

Fig. 3.2. Models according to the level of abstraction

Reverse engineering and modelling techniques can be combined into tools as meant from the current literature. Figure 3.2 presents the correspondence between the type of models and the levels of abstraction. Declarative models are characterised as the most abstract models since none geometric information is included. Feature models consist of both geometric and feature-based information.

3.1.2 Reverse Engineering and Geometric Modelling

The 3D reconstruction is a huge field and of course there is a vast literature on the specific subject. The input of reverse engineering processes is the physical object of interest, while the output is its CAD model: either a surface representation or a volume representation. The conversion between surface and volume model can be done by mature algorithms such as voxelization and marching cube algorithms. Surface representations and fitting methods for reverse engineering are summarized in [6], [24]. In conventional reverse engineering processes, 1D (point cloud data) or 2D (range images) sampling data of the surface of interest is acquired by digitizing devices (CMM, laser scanner, etc.) or photographing devices (CCD camera, ICT, MRI, etc.), respectively.

Reverse engineering methods diverges from data measuring strategies. Commonly used data acquiring devices can be categorized into contact or non-contact devices. Contact type devices are generally more accurate but slow in data acquisition, and vice versa for non-contact type devices. According to whether the probe is held by operators, contact devices can be further divided into two classes: automatic devices and manually holding devices. The common drawback of contact devices is that they may deform or even damage the surface of the object being digitized because of the direct physical contact. Non-contact devices measure the point coordinates using distance measuring methods, such as laser scanner and sonar. The merit of non-contact distance measuring methods is high scanning speed. The accuracy of this type of digitizers is dependant on surface reflectance, the light paths to the sensor, and the material (transparent or semitransparent et cetera).

In engineering areas such as aerospace, automotives, shipbuilding and medicine, it is difficult to create a CAD model of an existing product that has a free-form surface or a sculptured surface. In these cases, reverse engineering is an efficient approach to significantly reduce the product development cycle [37]. It is often used in cases such as the following:

- Where a prototype of the final product has been modelled manually and therefore no CAD model of the prototype exists, e.g. clay model in automotive industry.
- Where a CAD is introduced in a company and all existing products must be modelled in order to have a fully digital archive. Particularly, the CAD model of a complex shaped part is modelled because it is difficult to create its CAD model directly.
- Where complex shaped parts must be inspected and therefore the reverse engineering model created will be compared to an existing CAD model.

In medical engineering a representative example of reverse engineering is the customized artificial joint design [19], where in order to meet the requirements and to reduce the production cycle and cost, a method is presented to generate the complex surface of an artificial knee joint by co-ordinate measuring machine from the normative prosthesis, and form the model data base. The method gets the better data points among point cloud data and then, the free-form surfaces are constructed from the point cloud data using the reverse engineering software —Surfacer. The solid CAD model of the artificial knee joint is created from the surfaces by extension, intersection. These models formed the data base of the prosthesis, in which we can select a suitable kind of artificial knee joint model to customize for the patient. It is only needed to change the local data of the corresponding CAD model to meet the different requirements of the patient.

In [32] is introduced the ByzantineCAD, a geometric parametric CAD system for the design of pierced jewellery. The ByzantineCAD is an automated parametric system where the design of a piece of jewellery is expressed by a collection of parameters and constraints and the user's participation in the design process is through the definition of the parameter values for reproducing traditional jewellery. The design of the traditional pierced jewellery is based on the voxel-oriented feature-based Computer Aided Design paradigm where a large complex pierced design is created by appropriate placing elementary structural elements. The final piece of jewellery is produced by applying a sequence of operations on a number of elementary solids. An algorithm for scaling pierced patterns and designs has been introduced to enlarge pierced figures without altering the size of the structural elements used to construct them. The reverse engineering process is focused on providing manually the elementary structural elements to the system instead of capturing these elements from existing artifacts and using them to reproduce the originals.

3.1.3 Reverse Engineering and Feature-Based Modelling

In the feature modelling field, object semantics are semantically represented for a specific application domain. In other words, a semantic feature is an application-oriented feature defined on geometric elements. There are three approaches for building a feature model:

- The design by feature approach creates the feature model of an object by composing the available features in a feature library,
- The feature recognition approach recognizes various features from a geometric model of an object according to the feature templates defined in a feature library,
- The feature conversion approach enables the definition of other feature models based on a feature model of a product already created. The new feature model corresponds to alternative views of the same product. Feature conversion is a technique that defines the basis for multiple-view feature modelling systems [35].

In order to create a feature model of an object from a point cloud, the embedded features must be recognized. These features are then used to constrain the fitting process. Feature recognition methodologies can be classified into two major categories: surface recognition and volume recognition. The main difference between these two categories is mainly due to the feature definition and object description in the recognition method. The features and object description in the surface recognition category are expressed in terms of a set of faces and edges. The graph-based method discussed by [20], syntax-based method proposed by [17], rule-based method discussed by [15] and neural network method raised by [25] are examples of this category. On the contrary, the features and object description are defined in terms of primitives in the volume recognition category. Typical examples are the convex hull algorithm by [16], hint-based reasoning method by [14] and curvature region approach by [31]. Although many different methods are proposed to recognize the features from an object description, only regular shaped objects can be handled [20], [31].

In [1] is discussed the issues of applying feature technology to reverse engineering technology of a mannequin. According to that, the feature model of a mannequin consists of the major features of the torso for garment design, and the features are recognised from the point cloud by comparing it with a generic feature model. Association is set up between the point cloud and the mannequin feature. Fitting the generic model to the point cloud yields the mannequin feature model of a specific person. This is achieved by optimizing the distance between the point cloud and the feature surface, subject to the continuity requirements. However the task of matching the critical points is done manually. Since surface fitting forms the shape of the human model it is hard to capture details of the mannequin and the process of surface fitting is time consuming.

In [36] a feature-based approach is presented of building a human model from a point could. The noisy points are removed and the orientation of the human model is adjusted. A feature based mesh algorithm is applied on the point cloud to construct the mesh surface of a human model. The semantic features of the human model are extracted from the surface. The advantages of the specific approach are the topology of the human models preserved, more details can be included in the feature human model, and the algorithm seems more efficient.

Apart from these, in [10] is presented the contribution of knowledge in reverse engineering problems. In particular, it is discussed the applicability of domain knowledge of standard shapes and relation ships to solve or improve reverse engineering problems. The problems considered are how to enforce known relationships when data fitting, how to extract features even in very noisy data, how to get better shape parameter estimates and how to infer data about unseen features. Even if the current work focuses on the reconstruction, it shows that the applicability of domain knowledge, in the general framework of the knowledge-based approach, plays a significant role in the reverse process. The paper explores techniques, made at Edinburgh University, to improve reverse engineering of objects from three-dimensional (3D) point data sets by applying constraints on feature relationships in manufactured objects and buildings in order to improve the recovery of object models, by applying general shape knowledge for recovery even when data is very noisy, sparse or incomplete. Many of these recovery problems require discovery of shape and position parameters that satisfy the knowledge-derived constraints. Evolutionary search methods can be used to do this search effectively.

The research from the University of Utah [33], [8] who have also been investigating constrained reconstruction of parts from 3D data sets, particularly parts with pocket profiles. They categorized the types of engineering knowledge as domain-specific and pragmatic, and functional constraints. They exploit this knowledge to select surface types and manufacturing actions. Thus, with some user assistance, planar features that bound pockets are found. The contour that is swept to form the pocket can then be found automatically. Shape and positional constraints are represented and solved in a manner similar to [10].

The research from the University of Cardiff [3], [4] exploits designed-in relationships to improve reconstruction. In their case, a much larger set of relationships was explored, and the constraints arising from the relationships were used to reduce the dimensionality of the reconstruction parameter space. A sequential numerical constrainment process is used, which allowed them to detect and automatically reject inconsistent constraints. A nice alternative to fitting tangential and blend surfaces was to parameterize swept 2D features, with the cross-section of the inter-surface join/blend as the 2D feature.

A special branch of reverse engineering is the reverse design where in [35] during the reverse design of free-form shapes, existing shapes or features can be extracted and be inserted into a model, or otherwise reused for the creation of a new design. Here it is essential to note that the existing features might not be designed as such, but are perceived as an entity by the designer. In addition, the designer might expect that the feature possesses parameters that he/she can control, whereas such parameters were never defined. An important aspect of reverse design is therefore the interactive assignment of complex controls to shapes or features. These controls are needed by the designer to achieve shape modifications, which could be very situation dependent. The interactive assignment can be dependent on, or expressed in terms of, for example, characteristic points or curves in the shape under construction or in any other existing shape. This referring to features which were not designed as such is one form of reverse design.

3.1.4 Reverse Engineering and Declarative Modelling

The design process in declarative modelling is iterative. Each cycle consists of executing the phases of description, generation, understanding, and then the designer has to reconsider the initial description. After the execution of successive cycles the system must converge towards a set of alternative geometric solutions that are closer to the designer requirements. The process stops when the designer estimates that has achieved the goal. The iterative process can be represented by a spiral where in each successive iteration a set of geometric solutions are produced from a description which has been modified in order the solutions to converge to the most recent description. Figure 3.3 illustrates the iterative design process within the declarative modelling framework.

The concept of iterations has been discussed from many researchers. [9] speaks about the concept of outline, [18] presents the spiral diagram of problem solving, [30] organizes in cycles the sequence description, generation, forms, and [7] proposes also various working methods within the framework of the iterative design process.

The design process is considered as decomposable in several successive distinct stages pointing to the initial goals [21]. The design process can be viewed as a succession of transformations between descriptions. These descriptions can be classified to a representation hierarchy of most abstract to most concrete on levels of abstraction. In the framework of the declarative modelling, the design process transforms a declarative description into a set of alternative geometric solutions. Thus, two distinct levels of abstraction are presented the declarative, which represents the most abstract, and the geometric, which represents the most concrete.

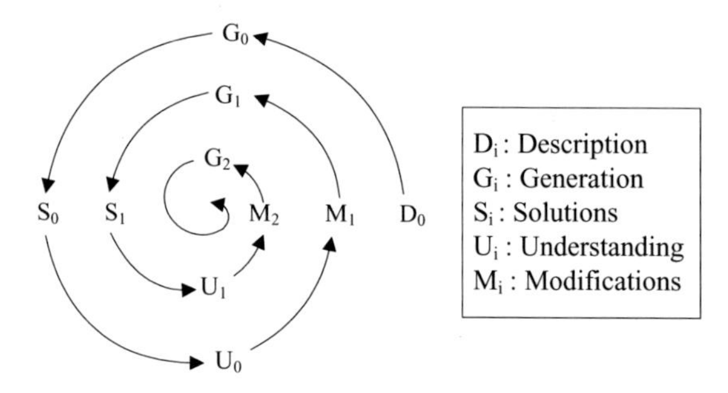

Fig. 3.3. The iterative design process

Generally speaking, the design process follows a walk from the general to specific [21]. In the majority of cases we start from an initial idea and arrive to details stage by stage. The levels of detail represent the hierarchical top-down or bottom-up approach of the design process. In the framework of the declarative modelling, the levels of detail are represented on the decomposition tree of the declarative description.

The evolutionary character of the design process is traced by the axis of time, the levels of abstraction and the levels of detail [22]. The design process is defined as a succession of similar stages for reaching the final model and can be represented as a spiral in a three-dimensional space. An initial idea serves as a starting point. As time evolves the level of detail increases while the level of abstraction decreases until the achievement of the final model. Figure 3.4 illustrates the evolution of the design process.

Reverse engineering in the declarative modelling framework acquires a geometric solution which can be modified by the designer and results a new declarative description to the next iteration of the declarative conception cycle.

The XMultiFormes project [27], [28] is a previous work that tries to integrate the two modellers by using a special interface system to ensure that there is full and complete transfer of information between the declarative and a traditional geometric modeller. This system is composed of four sub-processes, each of which is responsible for one aspect of the information transfer.

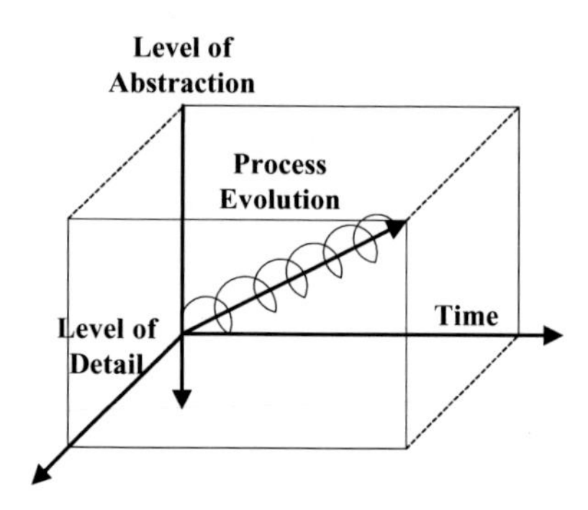

Fig. 3.4. Evolution of the design process

The geometric convection process translates the geometric representation generated by MultiFormes into one that is more suited to interactive modelling. The principal geometrical entity used by solution generator of MultiFormes is a closed parallelepiped which is formed by six connected surfaces of Bézier and is converted into three linked lists [29]. At the lowest level is a list of vertices which are used to implement a set of geometric primitives, the second level of representation. At the third level is the compound primitive of MultiFormes. At the highest level is the object entity, which is constructed from either a set of primitives or other objects.

The labelling system is responsible for capturing non-geometric information, which is implied in the declarative description. The principle source of non-geometric information are the sub-scene names in the declarative description. These manes typically contain an explicit high-level description of the meaning of the sub-scene, and are maintained by the labelling system as a set of properties called labels. After a geometric representation has been created a special process is used to traverse the hierarchical description of the scene and to match decomposed sub-scene names to labels. This process requires the cooperation of the user since the label list must be adapted to the set of sub-scene names.

The geometric-to-declarative representation conversion process converts a geometric instance to declarative description by identifying all objects without parent, determining if the bounding boxes overlap and, if so, decompose the hierarchical objects, generating a scene hierarchy using a binary sub-scene agglomeration and describing each sub-scene and their relationships. The scene inclusion process provides a means for the inclusion of previously generated scenes in a declarative description. The designer can associate a sub-scene in the declarative description with an existence scene or a list of existing scenes.

Sellinger connects successfully a traditional geometric with a declarative modeller, gives special emphasis on retrieving the appropriate knowledge from the designer and gives special attention on man machine interaction.. However, the XMultiFormes project does not incorporates any database management system, which means that the designer is obligated to execute the whole design process at once, and can not store the most desired geometric solutions for further manipulation. The lack of a database management system also affects the update on the relationships since the available relationships are hard coded obviously, and inhibits the designer to work on different domain other than the architectural design of buildings.

The inclusion process of XMultiFormes provides to the designer the only ability to modify declaratively the scene by including a sub-scene in the declarative description with an existence scene which makes the process inflexible for further modifications. Besides, the geometric to declarative representation process requires in each iteration computation time and the quality of the new declarative description, which is produced, is not evaluated since Sellinger does not present any convergence of the geometric solutions.

Generally speaking, the cooperative computer aided design paradigm (CCAD) was used by Sellinger as a framework for the integration of the two modellers. This paradigm was originally developed to provide an interactive generative geometric modelling system with a cyclic design path and it is well adapted to the declarative modelling. The CCAD paradigm is based on the assumption than the system can not create perfect geometric solutions, but superior designs can be generated by allowing the designer to guide the system through successive rounds.

3.2 Integration of the Two Models

Geometric and declarative models have useful aspects but are also in some way limited. A declarative model corresponds to a set of alternative geometric models and a geometric model corresponds to a set of alternative declarative models. Thus, the correspondence of the two models is not univocal. It must be pointed out that one model complements the other one in a way that many problems of the one model are solved by the other model. For this reason the integration of the two models results in an extremely powerful design tool.

Figure 3.5 presents the type of the acquired knowledge relative to the level of abstraction. On high-level of abstraction the valuable knowledge is more declarative than geometric and on low-level of abstraction the declarative is less than the geometric specific knowledge. On an intermediate level of abstraction the acquired knowledge is a combination of declarative and traditional geometric knowledge [12].

A model, in order to become another type of model, is gradually transformed into a sequence of different levels of abstraction by a sequence of processing steps. The imperativeness of introducing an intermediate model derives from the fact that the nature of a declarative model differs from the nature of a traditional geometric model.

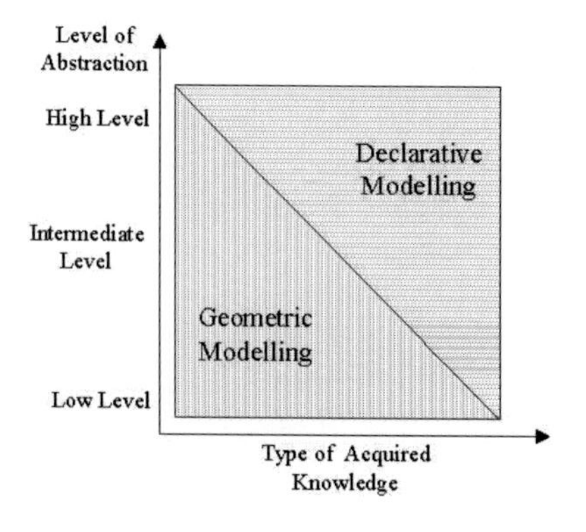

Fig. 3.5. Type of acquired knowledge according to level of abstraction

Our declarative model incorporates a decomposition tree representing the level of details, a set of object properties and a set of relations that connects the objects. A traditional geometric model incorporates a set of geometric information necessary for representing the objects of the scene. Thus, the intermediate model consists of declarative and geometric data of the scene that are connected properly.

Figure 3.6 shows schematically, the transformation of a geometric model into declarative via an intermediate model. The arrows that connect the different models show the reverse design within the declarative modelling framework.

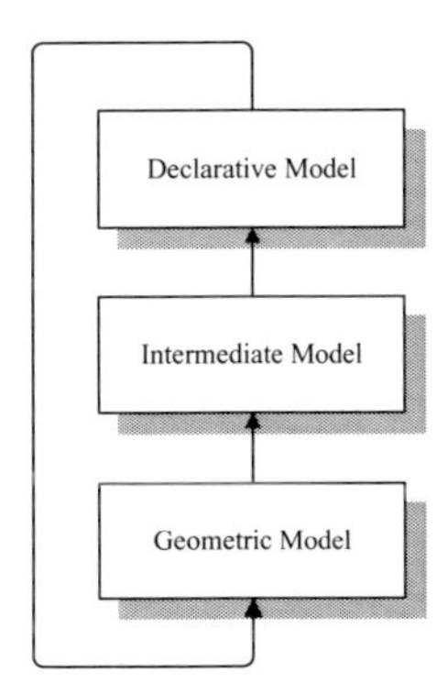

Fig. 3.6. The transformation of a geometric model into declarative

In the framework of the declarative modelling all geometric models, which have been produced from a specific declarative description, differ onto the values of the object relations and the values of the object properties. The different values of the properties cause an object with a new shape and position on the scene while the different relations cause a new arrangement of the scene.

3.3 Reconstruction Phase

The declarative conception cycle of MultiCAD system architecture can be extended to an iterative process by using a reconstruction phase [12] where the scene is semantically understood and refined by adding more detailed descriptions in successive rounds of declarative design process. In this case undesirable designs are cut from the set of solutions, the size of the solution set after each round of generation can be reduced and after a few iterations the designer gathers all promising solutions. Under the reconstruction phase an intermediate model is built in order to handle all the necessary information concerning the declarative and geometric side of the scene. The aim of the reconstruction phase is to receive a geometric model, provide a new declarative model enhanced with geometric constraints to the scene declarative phase and also permit the designer to change the geometry of the scene by modifying the geometric aspects of the objects. These changes are semantically checked and the intermediate model is updated.

Figure 3.7 presents the new declarative conception cycle by placing the new reconstruction phase in a UML use case diagram. The cycle starts with a declarative description, which produces a set of geometric solutions. The solutions are visualized and the designer selects the most desirable geometric solution. In the reconstruction phase, the designer can edit the geometric solution, a new declarative description is created, which contains the changes and a new cycle starts resulting to more promising solutions. The iterative process aims to produce scenes, which meet the designer requirements, after refinement. The designer can proclaim his requirements declaratively and geometrically.

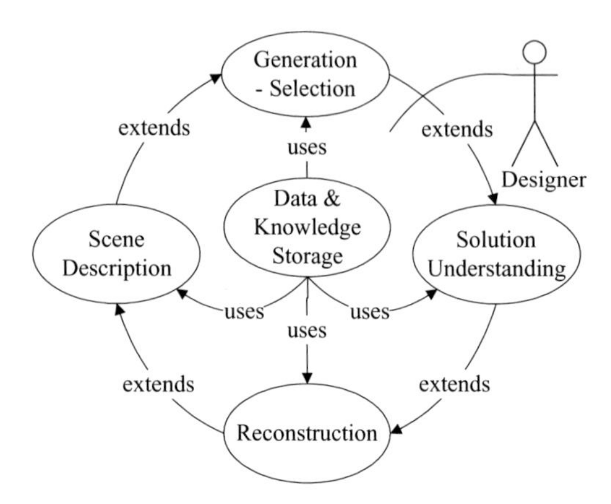

Fig. 3.7. The new declarative conception cycle

Due to the introduction of the reconstruction phase in the declarative conception cycle, the design methodology is extended and presented in the below sub-section.

3.4 Extended Design Methodology

The design methodology of the declarative modelling extends in order to include the reconstruction phase. The extended design methodology [13] starts with the description of the desired scene in terms of objects, relations and properties through an interface. A rule set and object set are built representing the designer requirements.

Initially, the object set consists of all objects of different level of abstraction, and the rule set consists of all relations, properties that the designer has declared during the declarative description phase. Based on that rule set, a set of geometrical solutions is produced by a solution generator. The solutions are visualized through a 3D viewer and the designer selects the most desirable solution, which can be edited.

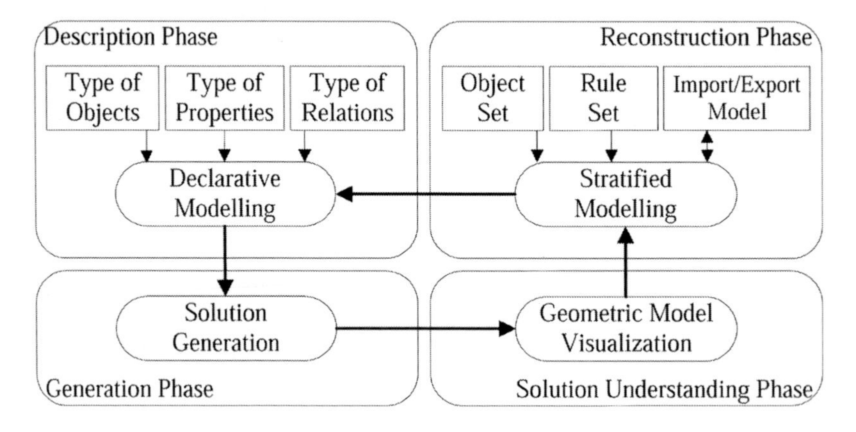

Fig. 3.8. Extended design methodology and modelling levels

The reconstruction phase is implemented through the RS-MultiCAD system, which receives the selected scene and converts into a stratified representation. The rule set and the object set can be edited by adding, deleting, and changing the objects, relations and properties of the scene. The designer can proclaim his requirements declaratively and geometrically during the reconstruction phase. A new declarative description is constructed, which contains the changes and a new MultiCAD cycle starts resulting in more promising solutions. The iterative process aims to produce scenes, which meet the requirements, after refinement. Figure 3.8 presents the extended design methodology and the modelling levels.

3.5 System Architecture

The high level architecture of the proposed RS-MultiCAD system can be seen as three main modules [12]. Our geometric model consists of information about the types of the objects and the geometric aspects of the objects. The input file provides the types of the objects along with the position and dimensions of the bounding box, which specify every object on the scene.

Generally speaking, the methods component consists of procedures and functions for extracting features and relations from the geometric model. The knowledge & concept component involves the knowledge database and concept database along with the mechanisms for retrieving the knowledge for the database. The control mechanism component incorporates all necessary mechanisms for building, handling and manipulating the intermediate model whenever the designer tries to alter the objects geometry of the scene. Figure 3.9 illustrates the general architecture of the RS-MultiCAD system.

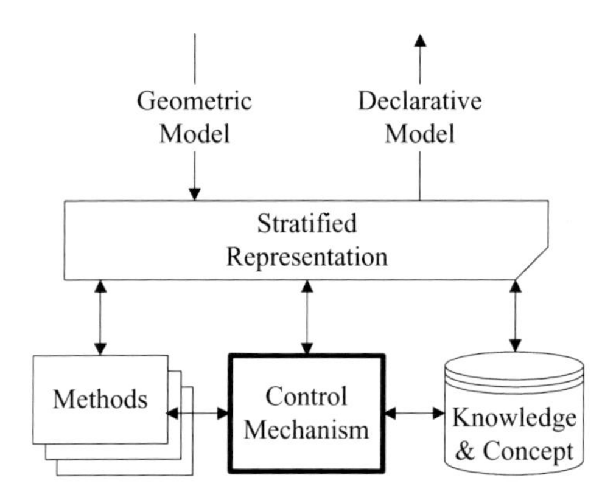

Fig. 3.9. General architecture of RS-MultiCAD system

The RS-MultiCAD knowledge-based system incorporates architectural domain-specific knowledge for constructing buildings. The system architecture is modular giving the possibility to further extensions. The system is based on five main modules from a high level of detail [13].

The import/export module is responsible for the communication with the databases supporting the input and output of geometric solution, the output of a new declarative description which comes from designer modifications, and finally the import and export of a geometrical model of different file format (DXF file format). The latter enhances the interoperability of the system since the designer can either import a design from another CAD system and produce alternative solutions or export the solution to other CAD system and continue the design process.

The extraction module applies all domain-specific relation and property types in order to extract all valid relations and properties of the objects from a selected solution. The extraction module is domain independent and facilitates the extension of knowledge and concept database since it parses the available knowledge from the databases. It must be pointed out that only the knowledge base is used for the extraction of the appropriate properties and relations from the scene and the concept database will contain all necessary concept representations.

The control module incorporates all necessary mechanisms for building, manipulating and updating the stratified representation. The stratified representation is dynamic and constructed from the designer selected solution with a top-down approach and mainly consists of declarative and geometric information. Declarative information can be summarized into object set and rule set. Geometric information deals with the geometry of each object that constitutes the scene. The control mechanism is event-driven and is responsible for the stratified representation to ensure the correct transition from one state to another. It handles the designer scene modifications examining their semantic correctness and properly updates the stratified representation by propagating the changes in a mixed way.

The explanation module provides valuable information about the system reasoning in cases where a scene modification violates the rule set. Finally, the RS-MultiCAD system incorporates a graphical user interface with a 3D editor in order to visualize the solutions and graphically receive the designer requests. Figure 3.10 illustrates the detailed RS-MultiCAD system architecture.

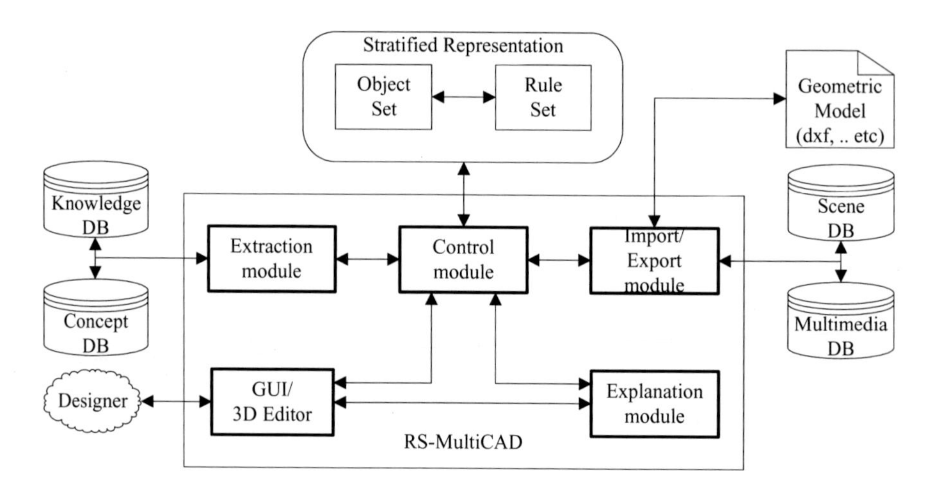

Fig. 3.10. Detailed system architecture of RS-MultiCAD system

The inner operation cycle of RS-MultiCAD incorporates all necessary mechanisms for converting a geometric model into the intermediate model and finally into the respective declarative model. The construction mechanism is responsible for receiving a MultiCAD geometric model and converting into the stratified representation. The extraction mechanism operates on the stratified representation in order to extract all appropriate relations and properties from the geometry of the objects that constitute the scene.

Besides, RS-MultiCAD incorporates an appropriate mechanism which permits the designer to perform modifications on the scene. Every designer modification affects the stratified representation. The system applies a specific propagation policy in order to properly handle the stratified representation and activates the extraction mechanism perspective to update the intermediate model. Finally, the intermediate model in converted into the resultant declarative description. Figure 3.11 presents the inner operation cycle of RS-MultiCAD.

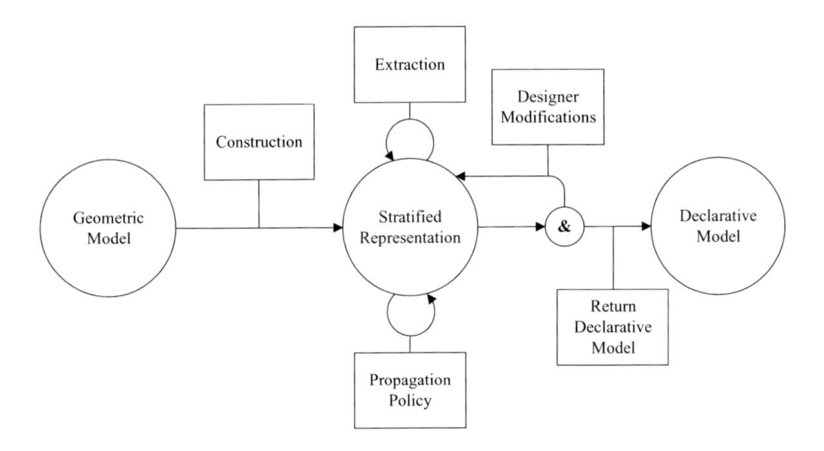

Fig. 3.11. The inner operation cycle of RS-MultiCAD

3.5.1 Data and Knowledge Storage

A brief description of the MultiCAD database is given in [22]. The RS-MultiCAD system in order to store a declarative description follows the guideline which is defined in [21]. A declarative description produces a set of alternative solutions that are stored in the multimedia database of MultiCAD.

Every geometric solution consists of a set of solution objects which in turn use geometric primitives which are expressed through geometric parameters. Thus, the multimedia database consists of a "Geometric Solution" table which is connected with the "Solution Object" and the "Solution Geometric Value" table. The "Solution Geometric Value" table incorporates the values of the parameters of the geometric primitive that describes every solution object of every geometric solution. Furthermore, the multimedia database has a direct connection with the knowledge database of MultiCAD. Additionally, the knowledge database incorporates two tables, the "Geometric Primitive" and the "Geometric Parameter" table. The "Geometric Primitive" table contains all primitive shapes which are supported by the system. The “Geometric Parameter” table includes all necessary parameters which characterize every primitive shape.

Every declarative description incorporates a set of objects. Every object corresponds to an object type. The declarative description also includes a set of object relations and a set of object properties. The object relation corresponds to a relation type indicating the relation between two objects. Every object property corresponds to a property type indicating the property which characterizes the object. Thus, the scene database contains the “Declarative Scene”, “Declarative Object”, “Object Relation” and “Object Property” table. The scene database in connected with the multimedia database indicating which geometric solutions correspond to each declarative description.

Apart from that, the scene database has a direct connection with the knowledge database of the MultiCAD. Additionally, the knowledge database incorporates three tables, the “Object Type”, “Relation Type” and the “Property Type” table. The knowledge database includes all object types, relation types and property types which are supported by the system. All object types are defined along with their primitive shape in the “Object Type” table.

The relation types are described through mathematical formulas in the “Relation Type” table. The mathematical formula is expressed through geometric parameters from the “Geometric Parameter” table. Finally, the property types are defined, within a range of a maximum and minimum value according to the domain in the “Property Type” table. The object property value corresponds to a declarative value which accrues from the subdivision of the range values into portions.

The knowledge database consists of all relation and property types which are supported by the system.

3.5.2 The Stratified Representation

The need of representing geometric and declarative information leads to an approach of using the stratified representation [26]. The stratified representation is an intermediate level model necessary for connecting the declarative with the geometric model, and embodies the two distinct interconnected layers of representation, the declarative layer which represents the scene description with the hierarchical decomposition and the geometric layer which encapsulates the geometric aspects of the objects [13].

The geometric layer of the stratified representation is based on the bounding box position and dimensions of each object which are expressed through the object pure geometric properties, along with any extra geometric information that can determine the shape of the object.

RS-MultiCAD inputs a geometric model which has been produced by the solution generator. That geometric model apart from the geometric information of all objects, that constitute the scene, provides their object type as well. The stratified representation is a dynamic semantic net with nodes and directed arrows. The node encapsulates declarative information of the object and the directed arrow reflects either the relation of the object with other objects or the object property or the level of detail. Every node is connected with a geometric node, which includes all relevant geometric aspects of the object, of the geometric layer. Figure 3.12 presents the basic structure schematically.

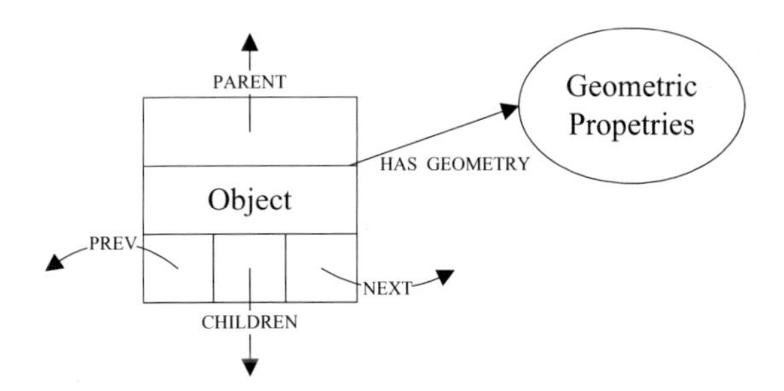

Fig. 3.12. The basic structure

Every node corresponds to an object and every arrow label indicates the relations of the node. The labels denote the following meanings:

The "parent" and "children" labels which connect nodes with same level of abstraction, different level of detail and represent the meronymic relations.

- The "next" and "previous" labels which connect nodes with the same level of abstraction and the same level of detail.
- The "has-geometry" label which connects nodes of different level of abstraction and represents the geometry of an object.
- The "has-topology" label which connects nodes of the same level of abstraction, indicating the topological relations among objects and represents the reflective and spatial relations.
- The "has-property" label is related to a node in order to indicate that the object has the specific property.

The construction of the stratified representation is a top-down process where the hierarchical decomposition is built based on the geometric information coming from the geometric model. For every object, a node is created on the geometric layer of the stratified representation. As long as all nodes have been created, the pure geometric properties lead to the hierarchical decomposition by creating interconnected nodes on the declarative layer of the representation. In Figure 3.13 appears a typical stratified representation.

The import of a geometric model needs a special process. The geometric model consists of geometric information of all objects that constitute the scene. RS-MultiCAD receives the geometric information of the objects and creates the respective nodes of the geometric layer of the stratified representation. The system creates one node of the declarative layer for every node of the geometric layer of the stratified representation. The nodes of the declarative layer constitute a linked list.

It must be pointed out that unlike the internal geometric model of MultiCAD, the external geometric model, which has been created by another commercial geometric modeller, may not contain any information about the object type and the name. In such a case, RS-MultiCAD set the object type of every node as "unknown" type.

As the time that the linked list has been created, RS-MultiCAD activates a specialized process in order to convert the linked list into the decomposition tree based on

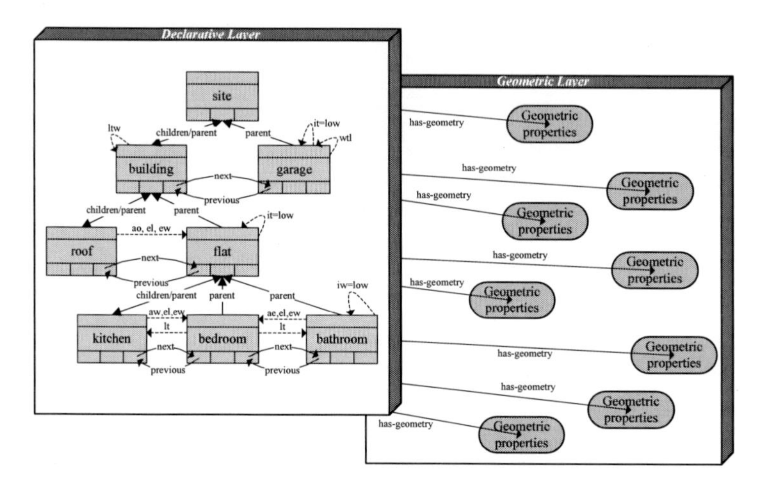

Fig. 3.13. A typical stratified representation

the geometric information of the respective nodes. The specialized process applies two transition operators:

- The "*consists-of*" transition operator which determines if an object O_i consists of another object O_j. The comparison is based on the position and the dimensions of the bounding box of the two objects. In other words, if the bounding box of the object O_i includes the bounding box of the object O_j then the "*consists-of*" transition operator is applicable and the object O_i becomes parent of the object O_j.
- The "*is-part-of*" transition operator which determines if an object O_i is part of another object O_j. The comparison is based again, on the position and the dimensions of the bounding box of the two objects. In other words, if the bounding box of the object O_i is included in the bounding box of the object O_j then the "*is-part-of*" transition operator is applicable, the object O_i becomes child of the object O_j.

The two transition operators are applied on every node of the linked list and when there is node where the "*is-part-of*" transition operator is not applicable with any other node, and the "*consists-of*" transition operator is applicable with all other nodes of the linked list, then it becomes parent of the rest nodes. In other words, the algorithm tries to find out which object bounding box includes all rest objects bounding boxes and simultaneously the former object bounding box is not included in none other objects bounding boxes.

The construction of the decomposition tree is based on the depth-first search and the process continues recursively for every node as any of the transition operators is applicable. For every level of detail that is created, the respective arrows of the nodes, that constitute the detail level, are redefined to point to the appropriate nodes. When both transition operators are not applicable to none node of a level of detail then the process backtracks to some node of a higher level of detail and the process goes on a different direction. The process stops when both transition operators are not applicable and all nodes have been processed. Figure 3.14 shows the decomposition tree after the process has been completed.

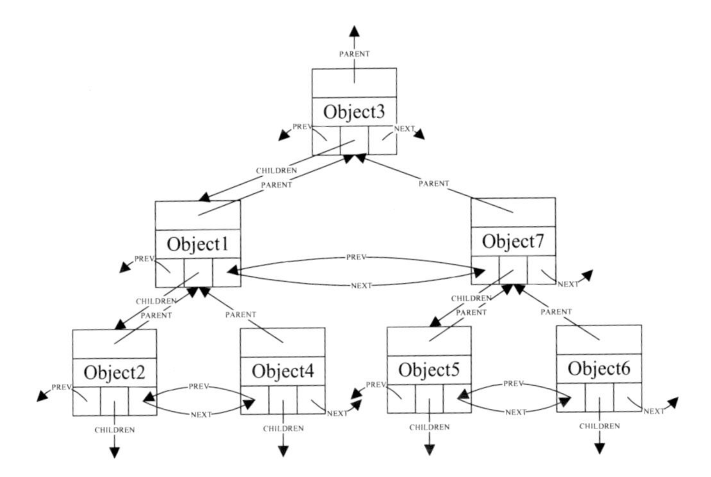

Fig. 3.14. The decomposition tree

At this point it must be pointed out that the algorithm is effective and all computations are quick. The algorithm has been tested with more than forty objects and it effectively operates on constructing the declarative decomposition tree and updating all respective arrow labels appropriately. Thereby, the intermediate model is appropriately constructed so it will be enriched, with semantic knowledge by exploiting the knowledge base and it will be correctly updated whenever the designer applies modifications on the scene.

3.5.3 Extraction of Relations and Properties

As soon as RS-MultiCAD has constructed the stratified representation, the next step is to extract all valid relations and properties which are accrued by applying all relation and property types of the knowledge database. Every node of the declarative layer of the stratified representation embodies two collection classes.

The first collection class contains all valid relations of the node and the second incorporates all properties along with their values. RS-MultiCAD uses a recursive algorithm in order to extract all relations and properties.

The system starts from the root node and uses a *preorder* way to traverse the decomposition tree. Using *preorder* the system visits, first of all, the root node, then the left sub-tree and finally the right sub-tree. On every node of the declarative layer of the stratified representation, the extraction module receives the geometric information of the object by following the "*has-geometry*" label. As the geometric data are known, the extraction module applies all stored properties in order to find out the current valid values.

The property types and their values are added to the property collection of the specific node. The system places the property type along with its declarative value, which has been properly calculated, to the property collection.

The extraction module continues applying all stored reflective relations and infers which are valid or not. The valid reflective relations are added to the relation collection of the specific node. The extraction module places the reflective relation type along with the object name to the relation collection. The extraction module also examines which spatial relations are valid. Preparative the system to compute which spatial relations are valid, it follows a specific tactic.

When the system visits a specific node in order to extract the spatial relations, it has to compare the geometry of the specific node with the geometry of the rest nodes, of the same level of detail, which share the same parent. The specific tactic uses the "*previous*" and "*next*" labels of each node in order to traverse all appropriate nodes of the same level of detail. The extraction module places the spatial relation type along with the related object name to the relation collection of the node. Figure 3.15 schematically presents which comparisons have to be made on the geometry of the nodes in order the system to extract the spatial relations of the specific decomposition tree.

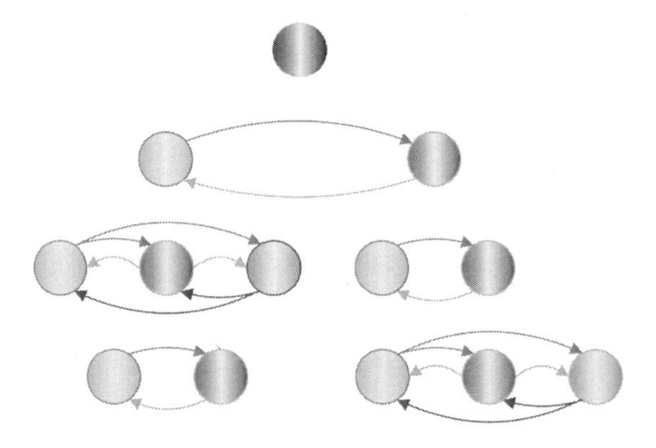

Fig. 3.15. The calculation of spatial relations

As soon as the process has been completed, the extraction module queries the scene database in order to find which are the relations and properties that were declared by the designer at the beginning. The extraction module then traverses the decomposition tree of the declarative layer in order to find out the respective relations and properties and mark them as "*designer requirement*". The rule set consists of these relations and properties.

3.5.4 Scene Modifications

The dynamic stratified model of RS-MultiCAD allows the designer to perform geometric and topological modifications on the scene [13]. As soon as the designer modifies the scene a special process starts. Every designer modification must be checked according to the rule set for its validity and if so the stratified representation must be properly updated in order to reflect the real state of the scene. RS-MultiCAD follows the "*generate-and-test*" method and provides two inference options according to designer modification which may or may not be activated:

- Check the modification according to the rule set. A modification is valid as long as none relation or property of the rule set is violated otherwise the modification is invalid and it is cancelled. If the designer decides not to check the modifications

according to the rule set, the control module performs a set of mandatory conditions ensuring the validity of the scene such as, none overlapping objects of the same level of abstraction, none object exceeding the overall scene limits, et cetera.
- Add pure geometric properties to the rule set that are inferred from the modifications. If the designer moves an object to a new position, pure geometric properties relative to move are added to the rule set.

The control module properly propagates the modification by updating the geometric layer of the stratified representation and activating the extraction module in order to recalculate all valid relations and properties. The control module assures that the transaction from one state to another one is valid since all relations, properties of the rule set are not violated and the changes are accepted while the new state of the stratified representation is valid. Otherwise, the explanation module is activated in order to record all violated relations, properties of the rule set and the control mechanism rolls the representation back to the previous state.

The modifications that can occur on the stratified model refer to abstract or leaf node and can be divided into two categories according to the geometric information that may be supplied by the designer. In particular, the declarative modifications are:

- Insert an abstract node by specifying its parent. The insertion of an abstract node in the stratified model can be done by specifying firstly an already existing node of the model as its parent and secondly the nodes that become children of the new abstract node. The result of such a change will affect the stratified representation since the object set changes.
- Delete a leaf/abstract node. The deletion of an abstract node will eliminate the subtree where the abstract node is root. The result of such a change will affect the object set and may affect the rule set as well. The stratified representation must be updated in order to reflect the current state of the scene.
- Set/unset relation, property. The designer changes the rule set by adding or deleting a relation or a property of a node.

Furthermore, the geometric modifications are:

- Move an object. The designer by providing the new position moves the object. The stratified representation must be updated since the move may affect the objects of higher and/or lower level of abstraction.
- Scale an object. The designer specifies the scale factor of the object.
- Resize object. The designer resizes the object by providing new values of the length, width and/or height of the object bounding box.
- Insert object. The insertion of a leaf node is carried out by specifying the geometric characteristics of the object.
- Alter the extra geometric characteristics of an object. In case where the shape of the object is complex, the designer can alter the extra geometric characteristics that define the shape of the object.

3.5.5 The Propagation Policy

The control module applies a specific propagation policy as soon as a modification occurs [13]. The propagation policy is necessary in order RS-MultiCAD system to recalculate and update the new position, new dimensions of the objects that are related to the object under modification. The main advantage of the propagation policy is the fact that the control module only updates the nodes that are related to the object under modification and leaves the rest nodes untouched. The main criterion of the propagation policy is based on the position of the object under modification into the decomposition tree of the stratified representation. An object under modification can correspond to a leaf node or an abstract node in the decomposition tree. Whenever a modification occurs on a leaf node then the propagation starts from recalculating and updating its brothers and continues to ancestors until the root node of the decomposition tree. On the left-hand side of the figure 3.16 is shown schematically that a modification on a leaf node (shaded area) affects its brothers and continues to ancestors (shaded area).

Whenever a modification occurs on an abstract node then the propagation starts from recalculating and updating its children and brothers and continues to children of brothers and to ancestors until the root node of the decomposition tree. On the right-hand side of the figure 3.16 is shown schematically that a modification on an abstract node (shaded area) affects first of all its children, then its brothers, children and continues to ancestors (shaded area).

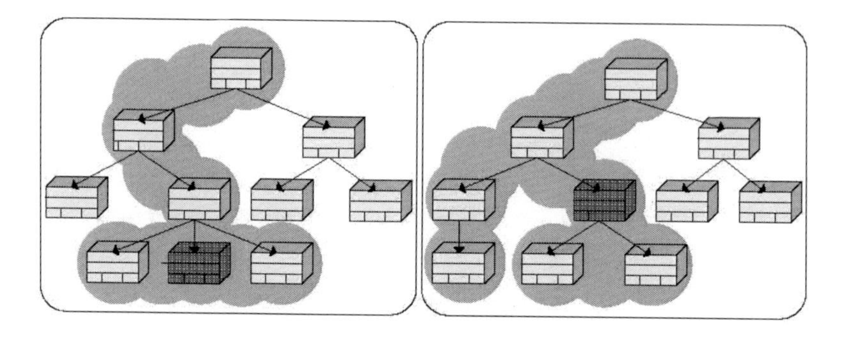

Fig. 3.16. The propagation policy

Let us suppose that an object of type "house" is decomposed into an object of type "bedroom" and an object of type "kitchen" which are related with a relation of type "adjacent north". A possible move, taking into consideration the aforementioned relation, of the object type "kitchen" causes the object of type "bedroom" to change its initial position, and afterwards causes the object of type "house" to change its position and dimensions respectively. At last, a possible move of the object type "house" causes the objects of type "bedroom" and "kitchen" to change their position.

The control module applies a specific propagation policy as soon as a modification occurs. Figure 3.17 presents the propagation policy. The object under modification and all related objects constitute a set. That set will be modified by the system. The control module calculates their new positions along with their new dimensions. Thus, RS-MultiCAD has already updated the geometric layer of the stratified representation only for the objects that belong to that set.

As the geometric layer of the stratified representation has been updated, RS-MultiCAD must update the declarative layer respectively. The process starts from the object under modification where the control module finds out the relations and the

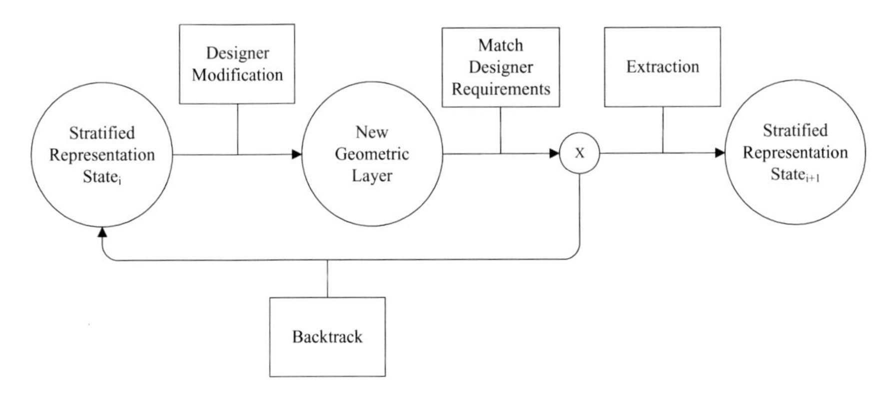

Fig. 3.17. The IDEF3 diagram of the propagation policy

properties which have been marked as "designer requirements" and belong to the collection classes. These relations and properties are applied on the new respective geometric node of the object under modification and if all these are valid, then the same process continues with the related objects, their descendants and accentors. This match operation is necessary in order RS-MultiCAD to examine if all designer requirements are still valid.

Otherwise the relations and properties which have been marked as "designer requirements" are not valid and they are sent to the explanation module. The process does not end and continues in order to find out all invalid relations and properties examining the related objects, their descendants and ancestors. The explanation module is activated informing the designer about the invalid relations and properties.

Whenever a modification occurs without taking into consideration the rule set, the control module through the propagation policy updates both layers of the object under modification, its descendants and ancestors without applying the match operation.

3.5.6 The Unified Stratified Representation

The specific methodology permits the designer to also select a set of geometric models and the corresponding set of the declarative models is constructed [11], [2]. As soon as the designer has completed all modifications on the scenes RS-MultiCAD first of all unifies the partial declarative descriptions, which include all modifications. Every selected scene has its own object set and rule set. The system has to unify all partial rule sets in order to produce the unified rule set.

The spatial organization relations refer to the connection of two objects. If two objects are related with complementary spatial relations then the system adds to the unified rule set the disjunction of the relations. For example if $Object_1$ is adjacent south to $Object_2$ belongs to the rule set of a selected scene and $Object_1$ is adjacent east to $Object_2$ belongs to the rule set of another selected scene then the system adds to the unified rule set the following relation:

[($Object_1$ is adjacent south to $Object_2$)

V

($Object_1$ is adjacent east to $Object_2$)]

The reflective relations refer to dimension comparison of the same object. If a reflective relation belongs to the rule set of a scene and its complementary belongs to the rule set of another scene then the system adds to the unified rule set the disjunction of the reflective relations. For example if $Object_l$ is longer than wide belongs to the rule set of a selected scene and $Object_l$ is wider than long belongs to the rule set of another selected scene then the system adds to the unified rule set the following relation:

[($Object_l$ is longer than wide)
V
($Object_l$ is wider than long)]

The object properties refer to characteristics that describe the object. If an object property is set to different values in different selected scenes then the system adds to the unified rule set the disjunction of the object property values. For example if $Object_l$ length='medium' belongs to the rule set of a selected scene and $Object_l$ length='low' is wider than long belongs to the rule set of another selected scene then the system adds to the unified rule set the following relation:

[($Object_l$ length='medium')
V
($Object_l$ length='low')]

3.5.7 The Resultant Declarative Description

As soon as the designer has completed all modifications on the scene, RS-MultiCAD results in a new declarative description which includes all modifications required by MultiCAD in order to generate in the next iteration more promising solutions by reducing the initial solution space. The question that arises is which relations and properties must be included in the new declarative description? RS-MultiCAD replies to that question by providing two optional ways, manual and automated [13].

In particular, RS-MultiCAD in the manual way results in a new rule set that is based on the initial rule set along with the new relations and properties that have been changed by the designer. In this way, RS-MultiCAD offers the designer the possibility to drive the system to generate a solution space that is closer to his requirements.

Furthermore, the automated way is based on the generalization factor (GF). Every hierarchical decomposed tree is divided in distinct levels of detail. The generalization factor is related to levels of detail and its values vary from 1 to maximum tree depth. The rule set that accrues from the automated way is based on the initial rule set along with all designer modifications and also all pure geometric properties that are implied from the generalization factor. The nodes that provide their pure geometric properties to the rule set are the nodes that belong to the same and higher level of detail if and only if these nodes have descendants. If the designer set the generalization factor to the maximum tree depth all nodes of the decomposition tree provide their pure geometric characteristics to the resultant rule set.

Figure 3.18 schematically shows which pure geometric properties are included in the rule set according to generalization factor. If the generalization factor equals to 1, the pure geometric properties of the root node are included in the rule set. If the generalization factor equals to 3, the nodes that provide their pure geometric properties to

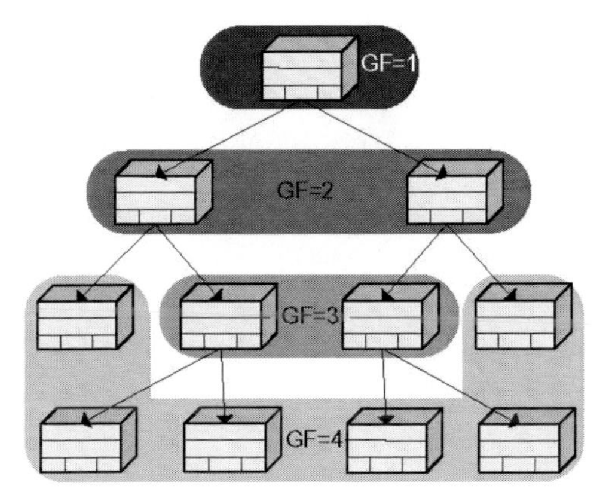

Fig. 3.18. The generalization factor

the rule set are the nodes that belong to the first, second and third level of detail except the nodes which have none child. As it is obvious, if the generalization factor equals to 4, the rule set is enriched with pure geometric properties of all nodes, which it will lead to only one geometric solution in the next iteration of MultiCAD.

3.6 Conclusions

The RS-MultiCAD system has moved on the following main directions:

- The scene is understood semantically. The RS-MultiCAD system receives a geometric model and in order to produce a new declarative scene description, constructs an intermediate model that contains geometric and declarative information about the objects that constitute the selected scene. In particular, the declarative information of the scene emerges from exploiting the knowledge base of MultiCAD declarative modeller. The current knowledge base of MultiCAD contains all the relevant knowledge on the types of objects, types of relations and types of properties which are involved in building design. Besides, the RS-MultiCAD system has been designed in a modular way in order to be able to understand any scene independently of the design.
- The declarative process becomes iterative and supported by machine. The RS-MultiCAD system implements the iterative nature of the declarative modelling by automating the process. A new declarative description emerges every time RS-MultiCAD receives a geometric model and supplies the declarative scene description of MultiCAD. Therefore, the designer liberated of the concern to redefine the declarative scene description in the next iteration.
- The scene can be modified during the reconstruction phase. The RS-MultiCAD system permits the designer to make all necessary modifications on a selected scene before the construction of the new declarative scene description in order the designer to alter his/her requirements. The modifications that can be applied are categorised according to the way the designer informs the system. Therefore, the designer can modify the geometry of an object directly by specifying new geometric values on one

hand, and on the other hand, he/she can alter the topology of the scene by performing modifications on the objects of the scene.
- The resultant declarative scene description can be affected by adapting appropriately the designer requirements. The RS-MultiCAD system maintains a rule set where the designer requirements are kept in terms of relations between objects and object properties. The designer can change the rule set by adding or deleting relations and/or properties. Therefore, by changing the rule set the resultant declarative scene description includes all relations and properties that belong to the rule set.

RS-MultiCAD can be compared with XMultiForms [27] since both systems have been designed to cope with the coupling of a declarative with a traditional geometric modeller. The two systems differ on the following points:

- The geometric model that is used for input is different in both systems. The XMultiFormes system receives a geometric model which has been produced by the MultiFormes declarative modeller. The MultiFormes solutions generator produces geometric models which are expressed in terms of a closed parallelepiped which is formed by six connected Bézier surfaces. The RS-MultiCAD system uses the geometric model which has been produced by MultiCAD and which contains all necessary information about the geometry of the objects that constitute the scene but also contains additional information about the shape and the type of the objects. Every object is expressed in terms of the position and the basic dimensions of its bounding box along with any extra geometric characteristics that describe the object.
- The RS-MultiCAD system constructs an intermediate model in order to capture both geometric and declarative information. The stratified model is characterised by flexibility on manipulating and absorbing the designer modifications. The XMultiForms system uses a labelling system in order to maintain declarative and geometric information in the same structure.
- In XMultiFormes system, the designer has the ability to change the scene geometrically unlike the RS-MultiCAD system provides both geometric and declarative scene modifications.
- In XMultiFormes system, the integration of declarative and geometric modeller is clear but does not integrate the iterative design process of declarative modelling. On the other hand, RS-MultiCAD facilitates and introduces the iterative process of declarative modelling supported by computer. The convergence of the set of the geometric models is achieved by the automated way, where the RS-MultiCAD system provides the appropriate geometric properties to the resultant declarative scene description, and by the manual way, where the designer adds further requirements to the resultant declarative scene description. It must be pointed out that the convergence of the set of the geometric models presents a steep slope in the automated way, unlike in the manual way, where the designer can decide, the slope can be smoother.

The RS-MultiCAD system improves the declarative scene modelling methodology, compared with the XMultiFormes system, by the following points:

- The RS-MultiCAD system reduces the cost of declarative modelling by permitting to first define a draft of the scene by declarative modelling and then refine the draft using an integrated geometric modeller. The designer can modify a selected scene and alter the rule set by adding relations and/or properties. Besides, the designer is able to modify the decomposition tree of the declarative description. Therefore, these modifications lead to a resultant declarative scene description which will corresponds to a set of geometric solutions that meet the designer requirements. Figure 3.19 presents a scenes of the system.

- The RS-MultiCAD system operates as an idiomatic navigator. The designer is allowed to apply modifications on the topology of the selected scene and the geometry of the objects that constitute the scene. The rule set always includes all designer requirements. Therefore, if these modifications do not violate the rule set, they lead to another geometric solution that belongs to the same solution space as the selected solution. Besides, if the modifications violate the rule set and the designer wishes to commit these modifications, they lead to another solution space and another resultant declarative scene description.

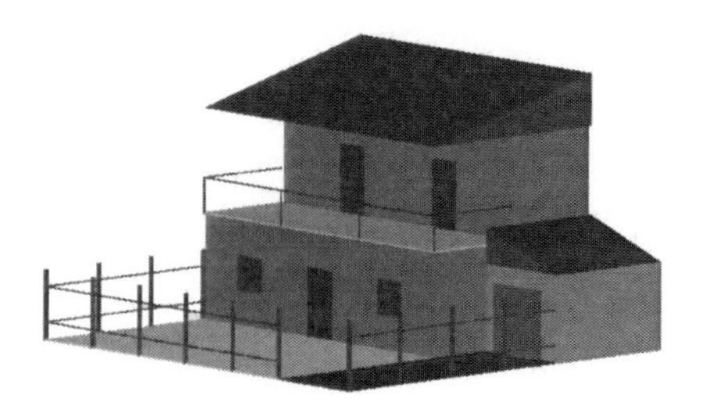

Fig. 3.19. Scenes after modification

- The RS-MultiCAD system gives the designer the possibility to subsume an initial geometric model, which is constructed by another traditional geometric modeller, in the declarative modelling methodology, and to benefit from its advantages. On the other hand, the designer through RS-MultiCAD system receives a geometric solution and elaborates the scene during the detailed design process with another traditional modeller.

References

1. Au, C.K., Yuen, M.M.F.: Feature-based reverse engineering of mannequin for garment design. Computer Aided Design 31(12), 751–759 (1999)
2. Bardis, G., Golfinopoulos, V., Miaoulis, G., Plemenos, D.: Abstract description refinement using incremental learning and scene reconstruction. In: 19th IEEE International Conference on Tools with Artificial Intelligence ICTAI 2007, Patras, Greece, vol. 2, pp. 345–348 (2007)
3. Benkó, P., Kós, G., Várady, T., Andor, L., Martin, R.R.: Constrained fitting in reverse engineering. Computer Aided Geometric Design 19, 173–205 (2002)
4. Benkó, P., Martin, R.R., Várady, T.: Algorithms for reverse engineering boundary representation models. Computer Aided Design 33(11), 839–851 (2001)

5. Chikofsky, E., James, H.: Cross II. Reverse engineering and design recovery: A taxonomy. IEEE Software, 13–17 (January 1990)
6. Chivate, P.N., Jablokow, A.G.: Review of surface representations and fitting for reverse engineering. Comput. Integrated Manufact. Syst. 8(3), 193–204 (1995)
7. Colin, C., Desmontils, E., Martin, J.Y., Mounier, J.P.: Working modes with a declarative modeler. Computer Networks ans ISDN Systems 30, 1875–1886 (1998)
8. de St Germain, H.J., Stark, S.R., Thompson, W.B., Henderson, T.C.: Constraint optimization and feature-based model construction for reverse engineering. In: Proc. ARPA Image Understand Workshop 1997(1997)
9. Desmontils, E.: Les modeleurs déclaratifs, rapport de recherche (Septembre 1995)
10. Robert, F.B.: Applying knowledge to reverse engineering problems. Computer Aided Design 36, 501–510 (2004)
11. Golfinopoulos, V., Bardis, G., Makris, D., Miaoulis, G., Plemenos, D.: Multiple scene understanding for declarative scene modelling. In: 10th 3IA, Athens, Greece, pp. 39–49 (2007) ISBN 2-914256-09-4
12. Golfinopoulos, V., Miaoulis, G., Plemenos, D.: A semantic approach for understanding and manipulating scenes. In: 3IA 2005, Limoges, France (2005)
13. Golfinopoulos, V., Stathopoulos, V., Miaoulis, G., Plemenos, D.: A knowledge-based reverse design system for declarative scene modeling. In: ICEIS 2006, Paphos, Cyprus (2006)
14. Han, J., Regli, W.C., Brooks, S.: Hint-based reasoning for feature recognition: status report. Computer Aided Design 30(13), 1003–1007 (1998)
15. Henderson, M.R.: Extraction of feature information from three-dimensional computer aided design data., PhD dissertation, Purdue University (1984)
16. Kim, Y.S.: Recognition of form features using convex decomposition. Computer Aided Design 24(9), 461–476 (1992)
17. Li, R.K.: A part-feature recognition system for rotational parts. International Journal of Computer Integrated Manufacturing 1(9), 1451–1475 (1988)
18. Liège, S.: Le modélisation déclarative incrémentale – Application a la conception urbaine., Thèse de doctorat, Nantes (Novembre 1996)
19. Yan-Ping, L., Cheng-Tao, W., Ke-Rong, D.: Reverse engineering in CAD model reconstruction of customized artificial joint. Medical Engineering & Physics 27, 189–193 (2005)
20. Little, G., Tuttle, R., Clark, D.E.R., Corney, J.: The Heriot–Watt feature finder: CIE 97 results. Computer Aided Design 30(13), 991–996 (1998)
21. Miaoulis, G.: Contribution à l'étude des Systèmes d'Information Multimédia et Intelligent dédiés à la Conception Déclarative Assistée par l'Ordinateur Le projet MultiCAD., PhD Thesis, University of Limoges, France (2002)
22. Miaoulis, G., Plemenos, D., Skourlas, C.: MultiCAD Database: Toward a unified data and knowledge representation for database scene modeling, 3IA 2000 Conference, Limoges, France (2000)
23. Peng, Q., Loftus, M.: Using image processing based on neural networks in reverse engineering. International Journal of Machine Tools & Manufacture 41, 625–640 (2001)
24. Petitjean, S.: A survey of methods for recovering quadrics in triangle meshes. ACM Comput. Surveys 34(2), 211–262 (2002)
25. Prabhakar, S., Henderson, M.R.: Automatic form-feature recognition using neural-network-based techniques on boundary representations of solid model. Computer Aided Design 24(7), 381–393 (1992)

26. Sagerer, G., Niewmann, H.: Semantic networks for understanding scenes. Plenum Press, New York (1997)
27. Sellinger, D.: Le modélisation géométrique déclarative interactive. Le couplage d'un modeleur déclaratif et d'un modeleur classique., Thèse, Université de Limoges, France (1998)
28. Sellinger, D.: Perspectives on the integration of geometric and declarative models for scene generation, Rapport de recherché MSI, Université de Limoges, France (1995)
29. Sellinger, D., Plemenos, D.: Interactive generative geometric modeling by geometric to declarative representation conversion. In: WSCG 1997, Plzen, Czech Republic, February 1997, pp. 504–513 (1997)
30. Siret, D.: Propositions pour une approche déclarative des ambiances dans le projet architectural – Application a l'ensoleillement., Thèse de doctorat, Nantes, France, Juin (1997)
31. Sonthi, R., Gadh, R.: MMCs and PPCs as constructs of curvature regions form feature determination. Computer Aided Design 30(13), 997–1001 (1998)
32. Stamati, V., Fudos, I.: A parametric feature-based CAD system for reproducing traditional pierced jewellery. Computer Aided Design 37, 431–449 (2005)
33. Thompson, W.B., Owen, J.C., de St Germain, H.J., Stark, S.R., Henderson, T.: Feature-based reverse engineering of mechanical parts. IEEE Trans. Robot. Automat. 15(1), 57–66 (1999)
34. Varady, T., Martin, R., Cox, J.: Reverse engineering of geometric models - an introduction. Comput. Aided Des. 29(4), 255–268 (1997)
35. Vergeest, J.S.M., Bronsvoort, W.F.: Towards reverse design of freeform shapes. In: WSCG 2005, Plzen, Czech Republic (2005)
36. Charlie, W.C.L., Terry, C.K.K., Matthew, Y.M.F.: From laser-scanned data to feature human model: a system based on fuzzy logic concept. Computer Aided Design 35, 241–253 (2003)
37. Yu, Z.: Research into the engineering application of reverse engineering technology. Journal of Materials Processing Technology 139, 472–475 (2003)

4

Intelligent Personalization in a Scene Modeling Environment

Georgios Bardis

Department of Computer Science, Technological Education Institute of Athens
gbardis@teiath.gr

Abstract. The current chapter presents an integrated approach towards the enrichment of a scene modeling information system with intelligent characteristics with respect to user preferences. The aim is to set a framework for this purpose covering the entire range of the scene modeling process, from the abstract definition to the final visual outcome. Necessary notions from preference modeling, decision analysis and machine learning are presented. Inherent difficulties of the current context are exposed and addressed. The architecture of an intelligent personalization module in the context of a scene modeling environment is defined and experimental results from its implementation and application are analyzed.

Keywords: Intelligent Personalization, User Profiles, Scene Modeling, Machine Learning, Multicriteria Decision Making, Preference Modeling.

4.1 Introduction

In his excellent short story *Library of Babel*, first published back in the 1940's, J.L.Borges presents an impressive metaphor, predictive of the modern day information era [3]. The location is a huge building of complicated architecture, consisting of narrow corridors, staircases and rooms all covered by shelves filled with books. The story is actually revolving around these books: given their size and the alphabet used, each contains only one of the numerous alternative sequences of letters that may be contained in such a volume and, as it is only natural, most of the times the entire book is incomprehensible or contains only miniscule fractions of reason. People living in this fictional library spend their lifetimes browsing through these books, seeking in vain for knowledge in their pages. An impressive aspect of the story is the implication that some of the most important books that have ever been written or, even, not written yet, already exist in the library simply because, they too, represent valid sequences of letters. However, due to the vast amounts of incomprehensible books, these really important volumes are extremely hard to discover. Moreover, the reader is left with the feeling that people who have been living and examining such content for years, might not be able to appreciate and gain from truly important books even if they had the chance to discover them.

The analogy with today's information overflow, although not intentional, is striking. The advent of steep technological advances in computational technology and

G. Miaoulis and D. Plemenos (Eds.): Intel. Scene Mod. Information Systems, SCI 181, pp. 89–119.
springerlink.com

telecommunications has made unprecedented amounts of information available under our fingertips or a few steps away in the form of websites, books, newspapers, magazines, TV shows, multimedia titles or publications. Nevertheless, one has to exercise judgment almost constantly, against diverse media and approaches, in order to be able to transform the incoming stream of information into an acceptable subset of knowledge or entertainment, not only in terms of size but also in terms of quality. Of course, most of the times, this elicitation is not – and cannot be – objective: one usually filters in the incoming information not necessarily according to strict intellectual ability but also according to subjective needs and preferences, ignoring parts that could be highly interesting for others.

4.2 Intelligent Personalization and Contributing Fields

Intelligent Personalization is a collective term used to represent a relatively wide range of techniques applied in a variety of contexts. Regardless of their diversity, these techniques share the common aim of automatic adaptation and conformance of an environment to personal preferences, with respect to interface and/or information provided. This task is far from trivial since human behavior can be inconsistent and, thus, unpredictable at different times and for diverse reasons. Either because humans and their environment are too complex to be accurately simulated and measured respectively, or because free will cannot be predicted (otherwise it should not really be considered free) it seems that, at least considering the present status of our knowledge and technical capacity, a deterministic non-human system cannot be expected to simulate human behavior in its entire range. This becomes also evident by the fact that Turing's Imitation Game has yet to be successfully passed by any artificial mechanism unlike its inventor's predictions [39].

Hence, the aim of intelligent personalization induces the inherent requirement for preference modeling, i.e. capture, maintenance and application of user preferences, under the assumption that such a model is feasible. Preference modeling has been the subject of extensive research over the years, thus bearing a strong theoretical foundation [45]. Depending on the context of application and the purpose of the personalization, it may vary significantly, ranging from the static adjustment of the user's interface to the intelligent adaptation of the environment as well as of the information accessed through this environment. Machine learning techniques are commonly adopted to capture and maintain different users' profiles that will be subsequently applied towards personalization in a scene modeling environment [1],[4],[35]. An alternative approach incorporates multicriteria decision support techniques in an effort to interpret and apply preferences [1].

In the following we examine the applicability of the most popular of the mechanisms originating from the aforementioned fields to an intelligent scene modeling information system. In order construct a framework appropriate for the current context, we concentrate on scene modeling, covering the whole range from the early-phase abstract description to the production of the final visualized result. The focus is set on the particularities of environments supporting this range in its entirety in regard to personalization. This ensures that the methodologies presented may be considered as a whole or individually and with respect to the different stages of either the process of design, in the case of visual information or the process of visualization, when discussing non-visual information.

The effort to endow a scene modeling environment with intelligent personalization characteristics requires the combination of methodologies and mechanisms from a wide range of fields. A visual representation of these interleaving fields appears in Figure 4.1.

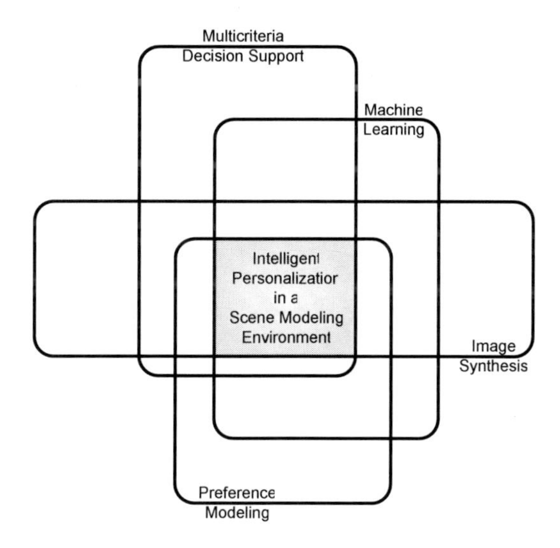

Fig. 4.1. Contributing Fields

Multicriteria Decision Support is a field of Decision Analysis which is concerned with the translation of the decision maker's preferences and expertise into a meaningful model which, when applied, will allow the selection of the most appropriate or preferable among a number of alternative available options. Several mechanisms have been developed to support this transfer from intuition to the corresponding computational representation [17], [22], [38]. The typical trade-off in this area is that between adequate user feedback and minimal overhead in terms of time and complexity regarding the acquisition of feedback on behalf of the user. Certain methods require pair-wise comparisons of the alternatives regarding each attribute affecting the final selection. Pair-wise comparisons represent one of the most intuitive, and, thus, widely accepted methods for obtaining user preferences among a collection of options. However, albeit simple and straightforward for the user, methods of this category become impractical when more than a few alternatives have to be evaluated with respect to a number of different attributes, since the number of comparisons that have to be performed becomes prohibitive. Other methods relax the need for pair-wise comparisons, compensating for it with the assumption of transitivity of preference. The latter assumes that if option a is preferred over option b and option b is preferred over option c then option a is, due to transitivity, preferable over option c for the specific user, which is not necessarily true in real world situations.

The issue of preference transitivity is typically addressed by Preference Modeling, a field concerned with the study of the intuitive notion of human preference with respect to alternative options. Relative to Decision Analysis in principle, yet this field bears its own sound mathematical foundations and incorporates contributions from cognitive science. Indifference or degrees of preference are issues of concern in order to compensate for inherent preference problems. One of them is the typical 'sugar' example: we may be indifferent between two spoonfuls of sugar differing by only one grain; however the accumulation of this indifference, grain by grain, may lead to clear preference in reality, without ever passing through a concrete threshold. The theoretical foundation of the field contributes to the clear understanding of any assumptions made in a specific context and their consequences with respect to the preference model applied.

Machine Learning is a field of Artificial Intelligence dedicated to the acquisition of knowledge or the simulation of knowledgeable behavior, on the basis of limited guidance in the form of rules or examples from different classes. Most methods in this area are based on learning by examples, hence, even if a set of rules has been predefined by the application area experts, problems connected with the quality and quantity of these examples arise. In particular, one of the main issues is the potential incompatibility between the simplicity sometimes implied by a finite set of examples and the increased complexity of the actual mechanism or function to be approximated. Another problem is that of over-fitting the examples, where the artificial mechanism, although successful with previously seen examples, is yet unable to behave correctly when new examples are presented to it. Last but not least, in cases where one or more of the classes are under-represented in the examples population, special care has to be taken to avoid consideration of these minority classes as noise on behalf of the mechanism. The effect of these problems may become apparent through the inability of the mechanism either to generalize its knowledge when presented with new examples or to acquire sufficient knowledge during its training.

Image synthesis is concerned with the processes and information connected with the visualization of data. The latter may originate from a variety of sources and may correspond to both visual or non-visual real world or fictional objects and concepts. In the current context we address both of these aspects and, hence, the synthesis of an artificial image is interpreted both as a way to simulate reality but also as an enhanced version of it, aiming to improve understanding and offer insight otherwise inaccessible from the stored information.

4.3 Preference Model

In order to extract a practical measure of the user's typically informal and immeasurable preference, we use the notion of *preference model* [32], [45].

4.3.1 Preference Structure

Formally, the set of *actions* is the set of candidates to be examined by the user as alternative options. In the present context, this may be – but not limited to – a set of visualized objects that comply with the abstract description submitted as input or a collection of data to be visualized by the system or a set of areas of a given image. The specifics of the object generation or visualization mechanisms are not directly relevant to the preference structure even if their operation may be connected with the latter. This is the case when, for example, the order of appearance is affected by preference or when the generation of alternatives is selective and not exhaustive.

User *Preference, Indifference* or *Incomparability* between any two actions $a_1, a_2 \in$ A is denoted as:

$$a_1 \,\mathrm{P}\, a_2 \quad a_1 \,\mathrm{I}\, a_2 \quad a_1 \,\mathrm{J}\, a_2$$

respectively where P, I, J represent relations over A. Formally

$$P = \{(a_1, a_2) \mid a_1, a_2 \in A\}, I = \{(a_1, a_2) \mid a_1, a_2 \in A\}, J = \{(a_1, a_2) \mid a_1, a_2 \in A\}$$

and, according to definition,

$$P \cap J = \varnothing, P \cap I = \varnothing, I \cap J = \varnothing$$

Indifference signifies equal preference whereas *Incomparability* signifies inability to compare due to lack of information or other reasons; therefore they represent two distinct concepts. Intuitively, these empty intersections imply that if one object is preferred over another then these objects are, obviously, comparable and the specific user is not indifferent between the two – hence the first and second empty intersection respectively. Moreover, a user being indifferent between two objects is obviously able to compare them thus leading to the third empty intersection.

In order to have a *valid preference structure* the aforementioned relations must have the following properties [45]:

$$\begin{gathered} \forall a_1, a_2 \in A: \\ a_1 \,\mathrm{P}\, a_2 \Rightarrow a_2 \,\not\mathrm{P}\, a_1 \\ a_1 \,\mathrm{I}\, a_1 \\ a_1 \,\mathrm{I}\, a_2 \Rightarrow a_2 \,\mathrm{I}\, a_1 \\ a_1 \,\not\mathrm{J}\, a_1 \\ a_1 \,\mathrm{J}\, a_2 \Rightarrow a_2 \,\mathrm{J}\, a_1 \end{gathered}$$

and, moreover

$$\forall a_1, a_2 \in A: (a_1 \,\mathrm{P}\, a_2) \veebar (a_2 \,\mathrm{P}\, a_1) \veebar (a_1 \,\mathrm{I}\, a_2) \veebar (a_1 \,\mathrm{J}\, a_2)$$

Notice that all the above imply that all objects are required to participate in exactly one of the aforementioned relations and, in the case of preference (as opposed to indifference or incomparability), a complete order is imposed. In particular, for any two objects from the set of options,

- one may be preferred to the other,
- they may be equally preferable or
- they may be incomparable.

It is worth noticing that the set of properties that has just been defined does not imply *transitivity* per se. It may be the case that a_1 P a_2 and a_2 P a_3 and, still, a_3 P a_1 without contradiction to the aforementioned properties. Hence, we may state that a valid preference structure does not necessarily require or imply transitivity.

4.3.2 User Preference as a Function

In the current section we use the term *user preference* for an unknown function $f{:}A \rightarrow \Re$ expressing, in direct analogy, the *numeric degrees* of a user's preference for the available options. We will assume that such a function exists and we will try to approximate it through a calculated function $p{:}A \rightarrow \Re$. This assumption is in conformance to the classical approach of *value function* or *utility function* (when cost is involved) of the Decision Analysis literature [32] and leads to important advantages but also bears a number of drawbacks. These mechanisms attach a numerical value to each scene (or part of a scene) which encapsulates all of its interesting properties and their impact on the user's preference; their aim is the construction of the function p that will approximate, to a reasonable extent, the user's intuitive function f. Notice that, by the definition of both functions, *all* available options are mapped to numbers. Formally, this includes the following definitions:

$$\forall a \in A : p(a) \in \Re$$
$$a_1 \, \mathrm{P} \, a_2 \Leftrightarrow p(a_1) > p(a_2)$$
$$a_1 \, \mathrm{I} \, a_2 \Leftrightarrow p(a_1) = p(a_2)$$

The consequences, in regard to the Preference Structure definitions presented above are the following:

- *No two solutions are incomparable.* Formally, $J = \varnothing$. Since any two solutions are represented by two numbers they can always be compared.
- *For any two solutions $a_1, a_2 \in A$ either one is preferable over the other or the user is indifferent between the two.* Formally,

$$\forall a_1, a_2 \in A : (a_1 \, \mathrm{P} \, a_2) \underline{\vee} (a_2 \, \mathrm{P} \, a_1) \underline{\vee} (a_1 \, \mathrm{I} \, a_2)$$

The reason is that since any two solutions are represented by two numbers they obey the *principle of trichotomy* of numbers stating that for any pair of numbers either they are equal or one is greater than the other. The above ensure that we use a valid Preference Structure. Moreover, they imply that *User preference and indifference are transitive.* Formally,

$$(a_1 \, \mathrm{P} \, a_2) \wedge (a_2 \, \mathrm{P} \, a_3) \Rightarrow (a_1 \, \mathrm{P} \, a_3)$$

$$(a_1 \, \mathrm{I} \, a_2) \wedge (a_2 \, \mathrm{I} \, a_3) \Rightarrow (a_1 \, \mathrm{I} \, a_3)$$

and also

$$(a_1 \, \mathrm{I} \, a_2) \wedge (a_2 \, \mathrm{P} \, a_3) \Rightarrow (a_1 \, \mathrm{P} \, a_3)$$

$$(a_1 \, \mathrm{P} \, a_2) \wedge (a_2 \, \mathrm{I} \, a_3) \Rightarrow (a_1 \, \mathrm{P} \, a_3)$$

The above are due to the fact that Preference and Indifference have been transformed into numeric relations. For example, the first clause of the first hypothesis implies that $p(s_1) > p(s_2)$ and the second clause implies that $p(s_2) > p(s_3)$ which, combined, imply that $p(s_1) > p(s_3)$. The argument is similar for Indifference as well as for the Preference/Indifference combination in any order.

The above set of inferences is a convenient consequence of the assumption of a preference function and, practically, one of the main reasons for its adoption. This advantage becomes concretely evident in situations with large numbers of options, where it is enough to calculate a value for each option thus eliminating pair-wise comparisons. Its main drawback is the inflexibility regarding incomparability which, in certain situations, lacks realism. However, as it will become apparent in the following section, the methodologies that may be applied in order to address incomparability typically require user feedback which may be infeasible to acquire when large numbers of alternative options have to be considered.

4.4 Multicriteria Decision Support

The typical context of application of Multicriteria Decision Support methodologies is one where a set of alternative options have to be evaluated by the Decision Maker against a set of diverse criteria that each contributes positively or negatively to the final outcome. The corresponding field is essentially part of the wider area of Decision Analysis that has been established during the 1960's as the formal procedure for the study of decision problems [18].

We can generally divide criteria used for the evaluation of the available options in two main categories: (a) *conscious* criteria that the user consciously employs in order to evaluate solutions and (b) *subconscious* criteria that contribute to the approval or rejection of a solution but are not realized by the user. The second category is covered by techniques from the area of Machine Learning, presented in a later section. Multicriteria Decision Support concentrates on the first category, where these criteria are well known and, therefore, can be adequately represented.

A major step towards the solution of a Multicriteria Decision problem is, having determined the set of criteria, to map these criteria to a set of *attributes*. These attributes have to be *common for all options* and will serve as the *quantifiers of the performance* of each option against these criteria. Essentially, each one of these criteria corresponds to an attribute that may be observed by the decision maker.

Several multicriteria decision support mechanisms have been proposed, including the Analytic Hierarchy Process [39] and SMART [17] – a variation of the Multiattribute Utility Theory [22]. A similar approach, requiring, however, minimal input on behalf of the Decision Maker, relies on standard weight assignment for the attributes [38], and is also of interest in the present context. All of these mechanisms are based on the *classical* assumption that every alternative option can be mapped to a numerical value, typically calculated as a weighted sum. An alternative direction is represented by the family of *outranking methods*, including the variations of ELECTRE [39] and PROMETHEE [4] mechanisms. A thorough overview of the Decision Analysis field can be found in [32].

The essential difference between the two directions of research is their flexibility regarding the *incomparability* of two options. The preference function approach circumvents this notion for reasons exposed in the previous section due to its structure: each alternative option is represented by a number and since any two numbers are comparable all options are mutually comparable. The outranking methods, on the other hand, directly address incomparability, a fact that yields several benefits towards realistic preference representation but may severely impede performance when the number of alternative options increases.

Both directions typically require, at some level, pair-wise comparisons of alternative options as part of their evaluation procedure. However, outranking methods heavily rely on the results of these comparisons in order to provide adequate feedback. Pair-wise comparison of all alternative options on behalf of the user is physically restricted by the human capability and usually lack of willingness to participate in such a tedious process. As a result, methodologies requiring it are applicable to problems with only a few alternatives to evaluate.

Apart from the increased number of alternative options, a large number of attributes to be considered also raises important issues for the personalization process. This phenomenon, also known as the *curse of dimensionality,* usually positions the problem outside the scope of decision analysis and is typically addressed using machine learning techniques, discussed in a subsequent section. In the following we briefly outline the available decision analysis techniques and highlight the most prominent representatives from each area.

4.4.1 Outranking Methods

Outranking methods were introduced in order to overcome certain difficulties that arise when attempting to apply weighted sum methodologies in specific contexts. In particular, these are cases where due to absence of adequate information or inability of the Decision Maker to select, two or more options may be rendered incomparable. These methods aim to address this fact and consider it as an acceptable phenomenon in the overall evaluation of the options.

Two major families of techniques have evolved in this direction: ELECTRE[1] methods [39] and PROMETHEE methods [4]. The majority of members of the former family concentrate on the degree of preference for each attribute separately, avoiding the integration of these degrees to a unique preference index for each option. Alternatively, preference is calculated taking into account the *intensity* of preference difference for any pair of options in every attribute and the frequency it occurs. Different variations employ different techniques in order to handle realistic situations such as uncertainty, inaccuracy or bias, using one or more discrete thresholds, functions playing the role of *veto* or additional pseudo-criteria and credibility indices. Moreover, certain representatives of the ELECTRE family aim at a complete order of the available options thus closely resembling the direction of preference function followed by the classical approach.

The family of PROMETHEE methods is also largely based on the assumption that it is unacceptable to combine an option's performance with respect to two or more attributes, in order to obtain its overall evaluation. Additional information is usually required by the method concerning the comparison of the attributes themselves and the varying degrees of preference within the context of a single attribute at a time. Certain representatives of this family relax the requirement for order, allowing selection of a subset of options without any order.

[1] Acronym from the French title: ELimination Et Choix Traduisant la REalité.

4.4.2 Weighted Sum Methodologies

This direction of research features some of the most popular representatives in the area of Multicriteria Decision Support. In the following we briefly present three representative approaches which may be applied in the context of a scene modeling environment offering varying feedback requirements. We hereby concentrate on their weight assignment policies leading to the definition of the preference function without focusing on other concerns, such as the stability analysis, which fall outside the scope of the current work. The representatives of concern for this discussion are: Analytic Hierarchy Process (AHP) [40] and its variants, Simple Multi-Attribute Rating Technique (SMART) [17] as well as Standard Weights Assignment policy [38]. AHP has been initially described in the early 1970's and has drawn considerable attention, being applied to diverse areas, implemented in commercial products [12] and triggering several interesting variations [6], [32]. The main structure employed by this family of methods is a pair-wise comparison matrix, used for the mutual comparison of all alternative attributes and options. The matrix operates under the axiom of non-transitivity of preference ratios as well as under the assumption of reciprocal judgments. Processing of the matrix leads to the complete ordering of both the attributes as well as the options. Due to the construction of the matrix contradictions may arise, revealed by an inconsistency ratio calculated upon the matrix content.

The SMART methodology relaxes the need for pair-wise comparison of the importance of the attributes requiring a single number for the relative weight of each attribute when all other attributes are muted for a specific option. These weights are normalized to yield the corresponding weight vector. The methodology, once again, is extended to the options through a weighted sum. However, the performance of each option against a specific attribute is calculated taking into account the middle value suggested by the user which offers a certain degree of adjustment of the distribution of the values, through the placement of the 'center' value anywhere within the acceptable range.

Standard Weight Assignment requires minimal user feedback in order to define the weight vector. The user has only to define the *order* of importance of the observed attributes thus declaring its *rank*. The weight vectors are automatically evaluated in one of four suggested alternative methods, each considering only the rank of each attribute and their total count.

4.5 Machine Learning

Machine Learning represents one of the dominant areas of modern Artificial Intelligence. Although the interpretation varies in the relevant literature, e.g. [24], [22], we may state that it represents *a set of computational techniques for the acquisition, elicitation and subsequent application of knowledge in regard to a partially or entirely unknown concept.*

Several categorizations exist with respect to the nature of the machine learning problems [7] that offer insight regarding the requirements and restrictions of the current context. One of these categorizations relies on the characteristics of the set of *examples* or *samples* used for the learning process. There are cases where a set of examples is available without any further characterization regarding the unknown concept, thus forming a problem of *unsupervised learning* where we expect the employed mechanism to recognize similar samples and group them, in the general case without prior knowledge regarding their number or their qualitative differences. In other cases, each example is accompanied by its classification, thus raising the issue of *supervised learning* in the sense that the mechanism is provided with information regarding the number of the different classes and the nature of their members. The problem of assigning each sample to one of a finite number of discrete classes is that of *classification,* whereas mapping each sample to a real numbered value is that of *regression.* Both categories of supervised learning problems may be reduced to approximating a multivariate function [22].

A crucial issue in supervised learning problems is the degree of relevance between the attributes used for the encoding of the samples, i.e. the *input* to the class or value suggested by the supervisor for the sample, i.e. the *expected output.* The relevant concern in the case of unsupervised learning is the requirement of interest on our behalf towards the automatic classification that takes place based on a given encoding of the samples. The task of selecting or constructing the appropriate attributes to represent the given examples, i.e. *feature selection* or *feature extraction,* is far from trivial and is usually aided by already existing domain knowledge for the concept to be learnt and/or dedicated mechanisms. In the following we will assume that such a relation between the input and the expected output exists for at least *some* of the attributes used for the description. We elaborate on this in a later section discussing the selected approach for the current context. Moreover, since the current context refers to the approval or rejection of generated solutions, or, in a wider interpretation, the characterization of each solution regarding user preferences, we concentrate on the most important mechanisms available for *supervised learning* for *classification.*

4.5.1 Traditional Machine Learning Mechanisms

According to a phrase usually ascribed to J.S.Denker, "neural networks are the second best way to do almost anything". The sentence implies that the first, i.e. best, way is to assign to one or more humans the tasks of understanding the different parameters of the problem, study its nature, gain adequate insight and implement a solution or a model closely resembling reality. Unfortunately, these tasks usually require a restrictive amount of time often exceeding that of a human lifetime. In other words, neural networks (NNs) have been considered the best way to exploit previous experience, i.e. existing examples of a concept, without necessarily delving into the nature of the concept itself in the form of rules or human understandable modeling. Formally, multilayer feed-forward neural networks have proven to be universal function approximators [18],[26]. Intuitively, this implies they may imitate any input-to-output mapping such as the evaluation of a scene on behalf of the user. Nevertheless, the quality and quantity requirements in regard to the data needed to achieve an adequate approximation may limit their applicability in the current context.

An extension to the phrase opening the previous paragraph, states *"...and genetic algorithms are the third"* [40]. Regardless whether this is a generally accepted statement or not, it implies that, at least, genetic algorithms (GAs) [18] are valued as a powerful machine learning tool, comprising the most important member of the wider family of evolutionary search techniques. The process of *natural selection* and *evolution* is artificially reproduced, based on the *survival (and reproduction) of the fittest* among a *population* of representatives comprising a set of solutions to a specific problem or a set of examples of a specific concept. Genetic problem solving typically deals with the evaluation and improvement of a population of candidate solutions or concept representatives. Intuitively, the algorithm's aim is to improve an initial solution population, by generating subsequent *generations,* of gradually improved performance. Ideally, after a number of generations, the representatives contained therein will be of high, although often not optimal, performance. In algorithmic terms, genetic techniques are *hill-climbing* algorithms, aiming to approach any local or global extrema.

The advent of Support Vector Machines (SVMs) [4] has gradually deprived neural networks some of their popularity. In contrast to the often empirical and application specific configuration of neural networks, support vector machines offer a firm and, up to an extent, uniform methodology with respect to different application areas. In particular, support vector machines form the implementation of a machine learning methodology based on Statistical Learning Theory [44] and represent a highly active contemporary research field. Although the architecture of a support vector machine is very similar to that of a neural network the construction method differs significantly. The learning process of a neural network aims to reduce the empirical risk of the network while maintaining a fixed network architecture, which implies a fixed confidence interval. Support Vector Machines, on the other hand, aim to reduce the confidence interval while maintaining a fixed empirical risk [44]. SVMs for classification are based on the idea of finding not *any* but the *optimal* boundary hyperplane for class separation. This optimality is achieved by selecting the hyperplane with equal and maximum distance from the closest representatives of each class, the *support vectors.* In a sense, the problem of separating efficiently all available examples is reduced to that separating efficiently the support vectors.

However, it is very often the case that the training examples are not linearly separable, at least when referring to a realistic problem, and, hence, the definition of a boundary hyperplane is not feasible in the space where the examples are defined. In order to overcome this, the examples are mapped to spaces of higher dimension in a non-linear way that ensures that classes will be linearly separable. The technique to achieve this is known as the kernel trick implemented by the kernel functions [Burbidge01].

4.5.2 Incremental Learning

The fundamental idea of the Boosting technique is the employment of *weak learners,* i.e. machine learning mechanisms of low performance, in order to create a committee of much higher performance than any of its members [42], [48]. [42] formally proved that any *weak learner*, i.e. an algorithm that may learn a class of concepts with accuracy slightly better than random guessing, can be recursively applied, making the error arbitrarily small, thus yielding a *strong learner.* The boosting technique practically applies the principles of incremental learning [11] in the sense that the overall machine learning mechanism is not of fixed size.

The technique takes advantage of the fact that a considerable amount of time during the training of a typical machine learning mechanism is spent for the fine tuning of its parameters in order to optimize its performance. As it is usually the case, this stage of adjustments contributes a relatively small error reduction, not proportional to the time required to achieve it. The Boosting approach, on the other hand, proposes the termination of the training of each one of its members as soon as its performance falls below a relatively high and quickly reached error margin. In a sense, each committee member is responsible of learning only a small *part* of the example set.

Another interpretation of incremental learning sets as the core issue addressed by the corresponding techniques that of *gradual availability of samples* to be used for training. This is usually due to external restrictions, posed by the application area of interest. It is also typically assumed that, whenever a new set of samples becomes available, re-training the mechanism with the entire set of samples is infeasible. In case this was not true, traditional learning mechanisms could be used, generally guaranteeing equal or higher efficiency for reasons explained below. However, training with the entire dataset may be infeasible due to time (numerous samples to use for training) or space (numerous samples to maintain in storage) restrictions that are also part of the application area characteristics.

The principal method used to address this problem is, again, the gradual growth of the learning mechanism both in terms of size and, more importantly, in terms of complexity. This growth is expected to tackle the corresponding complexity of the set of already seen samples when considered as a whole, as opposed to the reduced complexity represented by each subset of concurrently available samples. In a sense, this process of augmentation reflects the mechanism's "maturing" process with respect to increasing information availability.

Several approaches exist supporting incremental learning, usually relying onto one or more assumptions regarding the datasets. In particular, one of the major issues is that of adequate representation of each class, in each dataset. Dominant approaches of the field operate under the assumption of adequate class representation in every dataset whereas, whenever this is not assumed, a degradation of performance may be caused by the newly available samples [37]. Another relevant issue is that of the scope of each class within each dataset. In particular, samples in a newly available dataset may significantly extend the scope of one or more classes thus forcing the mechanism to reconsider its interpretation of class boundaries. The relevant assumption is that of restricted declination from a pre-defined class scope.

4.5.3 Imbalanced Datasets

Regardless of the learning mechanism type, i.e. traditional or incremental, the phenomenon of imbalance arises in contexts of classification problems where one or more classes are under-represented in the available dataset(s). This may be due to the nature of the area of application or the quality of the data. Due to the collective consideration of error rates, traditional classifiers tend to adapt to the over-represented classes, a fact caused by the error reduction this favoritism incurs. Moreover, classes of only a few representatives in the entire set may also be considered as noise. Regardless of the reasons for this imbalance, a series of methodologies have been proposed to address this problem [25],[47],[49] – [46] presents an overview.

One direction of techniques operates at the algorithm level through a series of mechanisms employing one or more of the following policies:

- Assignment of penalty for misclassification of the minority class samples.
- Application of alternative values for the classification threshold.
- Application of multiple classifiers, typically trained with different subsets, reaching consensus by majority voting.

The alternative direction proposes adaptations at the data level, through manipulation and transformations including:

- Generation of additional, synthetic samples for the minority class, by extrapolation on existing ones.
- Separation of the initial imbalanced set into balanced subsets where minority representatives are repeated whereas majority representatives are distributed.
- Replication of the samples of the minority class.
- Elimination of samples from the over-represented class.

4.5.4 Context Specific Issues

The current context poses restrictions that call for specialized machine learning approaches. Whereas the modeling process and scene representations are discussed later, the issues that arise with respect to the machine learning requirements can be discussed here.

In a general purpose scene modeling environment, covering the whole range of scene representations a user may submit an abstract scene description expecting from the system one or more valid geometric or image interpretations. The user is subsequently able to evaluate the proposed solutions, and either end the process or further refine the initial abstract description, thus restarting it. This core functionality of such an environment implies a number of technical issues that have to be addressed by a mechanism aiming to acquire user preferences. These issues may be summarized as follows:

- New solutions are produced and evaluated at distinct and distant times.
- Storing all produced solutions may be prohibitive in terms of space. This is due to the fact that an average abstract description may yield thousands of valid alternatives for each of numerous users.
- Retraining of mechanism using all solutions produced for a specific user may be prohibitive in terms of time. Even if we store large numbers of samples, it is not necessarily the case that they could be efficiently used for training.
- Different descriptions may lead to large differences in solution population and the user may be willing to evaluate only a few or all of the generated solutions. Thus, we need a mechanism able to adjust to varied degrees of complexity.
- The user may approve only a few solutions among numerous rejections. This implies the need for a learning mechanism able to handle imbalanced datasets.
- Small morphological differences may suggest large differences in user's preference. This fact practically prevents the adoption of mechanisms relying on geometrical similarity for the classification of solutions.

These characteristics inhibit the application of traditional machine learning mechanisms in the current context and point towards the direction of an incremental learning mechanism able to handle imbalanced datasets. Such a mechanism has been employed and tested for the transparent acquisition of user preferences, as described in the implementation discussed in a later section.

4.6 Intelligent Personalization in a Scene Modeling Environment

The general architecture of an intelligent personalization module, in the context of a scene modeling information system, has to integrate the effect user preferences at all possible levels of scene representation and interrelated tasks. Given the variety of activities these systems support, ranging from design or engineering to monitoring and detection, it is only possible to set a framework for this architecture and later present a concrete example in the context of one of these application areas. In the following, we present a model for the processes performed and data manipulated by such a module whereas in a subsequent section we present the implementation and behavior of such a module in the area of creativity support.

4.6.1 Scene Representations

The process of scene modeling covers a wide range of scene representations. The scene itself, regardless of the representation, may reflect information regarding a visual concept such as a city or an engine component, or a non-visual concept such as network traffic. The former is typical in the areas of design, architecture, engineering, etc. The latter refers to visualization of non-visual data for detection, analysis and understanding purposes. In order to cover the aforementioned process in its entirety, we consider in the following all major alternative scene representations of interest in the current context.

In general, the scene may start its life as an abstract description close to human intuition and, for this reason, lacking precision. This fact is even desirable in certain contexts where the lack of precision facilitates creative expression. This is true during early-phase design where a set of functionalities, ideas, constraints and other requirements are sought to be incorporated in the description of an object or environment. One of the dominant methods for abstract scene description is Declarative Modeling [27]. According to this methodology the scene is described as a set of abstract objects connected with abstract relations and bearing abstract properties. Declarative Modeling by Hierarchical Decomposition [34] enhances this description with intermediate levels of abstraction. At this level no exact values are defined for the properties, relations or the positions of the objects themselves. Practically, each property or relation represents a range of values, including one or more intervals.

The issue that has to be addressed at this level of description, with respect to personalization, is that of the *interpretation* of the abstract terms according to user preferences. In particular, a *large* house may represent a different total area for different users. Examples like this are valid for all terms used in the abstract description of an object or an environment. Similarly, in order to cover the case of non-visual data, a *large* number of customers, in a chart for sales performance, may represent different populations for different executives. It is important to realize, at this level, that the effort is not towards the definition of a ‘correct’ interpretation for each term because there is no single universal interpretation which would be appropriate for all users.

Rather than this, the aim is to adapt the practical (geometrical or other, depending on the case of scene) result of this interpretation to the needs and preferences of each user. Moreover, as a side-effect, this personalization could yield a possibly improved abstract description, tailored to the user's preferences. Such a description could, in turn, improve the efficiency of the mechanism responsible for the actual translation of the description into one or more visualized results. This level of personalization has been addressed in a series of works based on information from subsequent levels of scene representation [16], [28], [10].

An alternative scene description method incorporates predefined geometric objects, each adjusted by a set of attributes. At this level of scene representation the ambiguity has been eliminated and the issues that arise relevant to personalization are those connected with scene presentation and observation on behalf of the user. In particular, although the morphology of the scene is concretely defined, the aspects of the scene that may be of interest to the user are subject to customization. The outcome of personalization at this level may be alternative static views of the scene as well as trajectories of scene exploration according to preference criteria [36]. Moreover, given the geometric nature of this level of description, the presentation may be enriched by additional features known to improve the specific user's perception of the scene such as texture, color, etc. The result of the personalization, once again, would be increased efficiency of the supporting system, in the sense that the user may be alleviated from the task of tedious scene exploration in order to obtain in-depth understanding of its properties.

In order to achieve an integrated approach for intelligent personalization in a scene modeling environment, we also have to consider the level of scene representation as a pure image, which covers the lowest level in the abstraction hierarchy. Although the representation is different than that of the previous level, the aims of a personalization module are similar, in the sense that it could isolate areas of the scene that might be of interest for the user. Moreover, personalization may be interpreted as the assessment of the image, with respect to a number of user-specific criteria. Such a feature in this

Table 4.1. Personalization effect regarding scene abstraction levels

Scene Abstraction Level	*Personalization Effect: Data*	*Personalization Effect: Process*
High (e.g. Declarative Description)	Adaptive interpretation of value ranges	Increased availability of preferred features
	Provision of custom-built templates Intelligent Solution Production	
Medium (e.g. Geometric Models)	Automatic model evaluation Isolation of interesting objects and combinations Content enrichment	Presentation in descending order of calculated preference Adaptive intra-scene presentation
Low (e.g. Bitmaps)	Automatic image evaluation Isolation of interesting areas Content abstraction	Presentation in descending order of calculated preference Adaptive intra-scene presentation

context could relieve the user from the task of inspecting the scene altogether, in case this is desirable. Last but not least, at this level of possibly overflowing visual information, the scene could be abstracted, concealing non-informative characteristics, thus allowing the user to focus on its most interesting aspects according to his/her preferences.

Table 4.1 summarizes the effects of personalization with respect to each level of abstraction. The categorization of these effects distinguishes those influencing the information manipulated by the environment from those affecting the user's interaction with it. In certain cases this distinction is not clear, as in the case of intelligent solution generation which is equally relevant to the data as well as the process facets of a scene modeling environment.

4.6.2 Scene Modeling Process

In the current section we examine the scene modeling process and the interconnection of the different levels of representation presented above. The focus is set on the transition between subsequent levels of abstraction and the functionality which is necessary for the provision of intelligent personalization features. The common aim is that of improving the efficiency of such a system with respect to a number of measurable technical characteristics as well as immeasurable aspects of user satisfaction and ease of use of the corresponding environment.

In the general case, the scene is initially described at a high level of abstraction, in ambiguous terms. Given the flexibility of this description and the possibly large number of alternative interpretations, a mechanism is typically employed to generate a number of outcomes, all compliant with the initial description, yet of different degrees of interest for the user. The generation mechanism may vary, depending on the needs of the context and the willingness of the user to participate in the process.

4.6.2.1 Solution Generation: Constraint Satisfaction Techniques

In this case the initial description is considered as a system of constraints, expressed in mathematical form, restricting the values of the scene parameters. These parameters refer to the next level of scene representation, i.e. the geometric level. The mechanism is usually responsible for the generation of *all* alternative solutions, thus implementing an exhaustive search of the solution space. Personalization, in this case, may interfere in more than ways with the definition of the system of constraints before these are submitted to the generation mechanism.

In particular, based on a user's profile, the system may interpret the *declarative* relations and properties, included in the initial description, into specific ranges and parameter inequalities, not common for all users. In this case, the personalization mechanism practically reduces the degree of abstraction with the expected benefit of the restriction of the solution space. This is highly desirable in constraint satisfaction approaches since the restriction of the range of a single variable may lead to the elimination of large populations of solutions which, in turn, implies higher system throughput. Moreover, a personalization module could enrich the submitted declarative description with a number of additional declarative relations, properties or even objects, deemed appropriate for the specific user. Although the addition of objects may augment the solution space instead of restricting it, the enrichment of the description in general may lead to solutions closer to the specific user's preferences. This could become more evident in the case where the mechanism has managed to acquire certain subconscious criteria from previous user behavior. In such a case, these criteria could be incorporated in the description even if the user is not fully aware of them, at least in a formulated manner.

4.6.2.2 Solution Generation: Evolutionary Techniques

This approach considers the abstract scene description as the basis for the construction of an initial solution population which is by no means exhaustive of the solution space. Nevertheless, it is desirable that this population includes representatives of different areas of this space in an effort to ensure their representation in subsequent generations. This initial population and all that follow are evaluated and subsequently subjected to the typical evolutionary operations for the construction of the subsequent generations.

Personalization in this category of approaches may be represented by the evaluation function responsible for the assessment of each generation of solutions. At the trivial extreme of the range of personalization degrees, we could have the user manually evaluating each generation, thus guiding the evolutionary mechanism during the solution generation process. Such an approach represents maximal user participation and is typically not desirable on behalf of the user. Although the effort of the human evaluator might be similar, or even less, in comparison with the evaluation of the outcome of an exhaustive search, it presents the drawback of the interleaving of the generation process with the human intervention whereas, in the exhaustive search case, solution generation may take place off-line. On the other extreme, a user's profile could be incorporated to play the role previously held by the human evaluator. Such an approach, of course, requires prior acquisition of user preferences and construction of an applicable user profile, in contrast with the highly intuitive human evaluation. The main benefit of an automated approach in this case is the elimination of the requirement for real-time presence of the human evaluator during the generation process. Of course, the application of evolutionary methods may also benefit from the restriction of ranges or the enrichment of the abstract description in a manner similar to the one discussed above. However, the restriction of the solution space in this case may offer less obvious benefits due to the selective exploration performed by evolutionary techniques.

4.6.2.3 Solution Visualization

Once the concrete (geometric or other) representation of solutions is available, they may be visualized and inspected by the user. At this stage, a personalization mechanism may improve the user's interaction with the system, resolving a variety of issues arising at this stage. An example of such an improvement would be the automatic evaluation of solutions. This evaluation might lead to the rejection of a possibly large part of the solution population. In combination with this, or in the case where this is not desirable, solutions could be presented in *descending order* of preference. In this manner, the user could inspect only a subset of the generated solutions, knowing that they represent the best candidates for the fulfillment of the submitted description, in accordance to his/her profile. This can be especially useful in a context where the exhaustive search of the solution space may yield numerous valid representatives.

The visualization of each solution separately could also benefit from the application of each user's preferences. In particular, the presentation of the scene – through the currently visualized solution – may focus on interesting areas. In the case of three-dimensional environments, this presentation may follow a trajectory for scene exploration guided by the specific user's profile. It is interesting to remember that the scene may not correspond to a visual object, being a graph or other construction presenting non-visual information in a visual manner. Scene exploration may be further refined by user queries and requests at this stage. Notice that the requirement for user's presence and interaction with the system is feasible since the latter are given during the visualization and inspection stage.

4.6.2.4 Scene Modeling Environment

The notion of an intelligently adapting scene modeling environment may be augmented to include elements of the user's interaction with it. For example, the profile information may extend to features connected with the manipulation of the scene, in any of its aforementioned forms, and not only with its content. This may imply adaptation of the user interface according to preferences or adjustment of configuration details, possibly transparent for the user. An example of the former is a dynamic environment for the syntax of an abstract description providing instant access to preferred declarative elements or even automatically custom-built templates, facilitating the initial construction of the description. In regard to the latter, the anticipation of the evaluation and visualization of specific representatives or areas of a single scene, expected due to profile information, might be accelerated by employing, for example, optimized storage access for the specific representatives or early initiation of required rendering of the interesting areas. Although such features would be less obvious to the user, they could highly contribute to the improvement of the efficiency of such an environment. In a sense, an environment bearing intelligent personalization characteristics should be adaptable at all levels of visual interaction, either the latter is connected with the content, or the means used to access this content.

4.6.3 Preferences Acquisition

The above analysis has concentrated on the exploitation of user preferences towards the provision of an adaptive environment for scene modeling. The underlying assumption has been the existence of a mechanism responsible for the acquisition and application of these preferences. The general areas offering methodologies for this purpose have already been discussed in the previous. At this point we concentrate on the needs of such an environment and the general architecture of a module responsible for this task.

More than one parameters affect the choice of preference acquisition mechanism and its configuration. A principal parameter is the one directly connected with the user, namely that concerning his/her participation to this process. In particular, at one end of the range we find the entirely transparent approach, where user preferences are collected and formulated without any explicit user intervention apart from regular system use. This seems to be the 'user-friendlier' approach presenting a series of advantages and drawbacks that have to be exclusively considered against the planned area of application. The main advantage is that of transparency and the minimal overhead it suggests. As a result, the efficiency of the system will gradually improve and the potential of this improvement is equally available to all users. The main drawback is that of minimal availability during the initial sessions of system use. Moreover, the degree of user involvement with the system and the variety, or lack of it, of the tasks performed by him/her, both in regard to the content of the scene and to the features used, would greatly influence the quality of the corresponding user profile information. In any case, the geometric or other representations of the scene subject to user evaluation have to be encoded before they may be exploitable by a learning mechanism.

4.6.3.1 Solution Encoding for Preferences Acquisition

In order to allow a learning mechanism to obtain information towards profile construction, the scene representations evaluated by the user have to be in a form appropriate to be used as training examples. The main issue to be addressed is that of feature extraction since not all scene elements may be of interest in regard to user preferences. This task may be performed transparently, using unsupervised learning techniques in order to classify solutions according to what seem to be their most important properties. In this case an initial encoding may be required, in order to construct a set of uniform vectors originating from geometric or other solution representations containing varied objects and bearing different properties. The set of different dimensions of this initial encoding could be as wide as the system designer requires it, in the sense that this choice practically represents a trade-off: a large number of dimensions could make the task of a subsequent learning mechanism infeasible whereas a small number of dimensions may sacrifice completeness. Once this encoding has taken place, solutions may be automatically classified according to similarity. Representatives of each class may be subsequently evaluated by the user and used, alone or together with other class representatives, for training of the learning mechanism. Unfortunately, similarity in terms of morphology or other scene properties may not necessarily imply similarity in preference: miniscule morphological differences may cover the intuitive distance between approval and rejection on behalf of the user. Nevertheless, depending on the intricacies of the application area and the size of the problem, such an approach may represent a feasible alternative.

A variant scheme is the employment of pre-defined attributes, directly relevant to user preferences parameters, to construct an encoded version of only the interesting properties of the solutions. Additional information could be incorporated to these vectors in case these representations have originated from an abstract description. In contrast with the previous approach, which aims to encode *all* scene properties, this approach operates under the assumption that the pre-defined attributes completely identify the features of the scene which are relevant to user preferences.

4.6.3.2 Preferences Acquisition via Incremental Learning

When no additional feedback is provided by the user the system has to rely solely on the submitted and processed content and the executed actions. Mechanisms able to handle this kind of information originate from the area of machine learning and we may state the issue is essentially reduced to a problem of incremental learning. This is due to the fact that information is not available in its entirety from the beginning; the additional examples of content and behavior have to be incorporated to the knowledge already contained in the corresponding profile. This need is amplified by the fact that, due to the possibly large size of the content, it is practically infeasible to store and maintain data produced or processed during previous user sessions. Moreover, a large number of users may further increase the required storage. Therefore, powerful methodologies requiring the entire dataset for the training, or retraining, of the learning mechanism are not applicable in the current context.

Even under the light of incremental learning the current context poses further restrictions due to the quality of the newly available feedback. In particular, examples from the current session most probably will cover only a subset of the user's interests and therefore may be minimally supported by already acquired knowledge. This leads to performance degradation, especially at the initial stages of the mechanism's training. Last but not least, even in the basic case where the user simply approves or rejects examples, we may encounter the phenomenon of imbalance due to the approval of only a small fraction of these examples, which may be the case in a realistic environment. Therefore, additional care has to be taken in order to address this imbalance.

Transparent acquisition of user's preferences can be mainly based on the user's evaluation of presented examples. The latter may originate from an abstract description in which case, the approved representatives reveal aspects of the user's interpretation of the ambiguous abstract description terms. This knowledge may not be restricted to these terms themselves, but may also cover additional relations and properties not supported by the description repertoire. Such an approach offers the potential of an improved interpretation of the abstract description, enhanced by additional expressions restricting the range of the scene parameters. This could result in the restriction of the solution space corresponding to the submitted description and the relevant efficiency gain.

Evaluated examples also allow the interpretation of a user's preference with respect to the importance of a number of observed morphological attributes. Tendencies or extreme variance for the values of these attributes may signify the difference between interest and indifference by the user. Such a mechanism could examine criteria not necessarily consciously employed by the user. This is dependent upon the encoding scheme adopted and the wealth of information maintained in the produced vector compared with the original scene representation.

4.6.3.3 User-Assisted Acquisition of Preferences

In cases where it is acceptable to require additional feedback on behalf of the user, beyond regular system use, a variety of methodologies are applicable. We may distinguish the additional user feedback either as a pre-processing step, required only once for each user, at the initial session in the environment, or as an extension of the scene modeling functionality of the system.

In the former case, the user may be requested to answer a series of questions regarding preferences with respect to the content as well as the environment itself. The form of the questions may range from the formulated approaches of decision analysis techniques aiming to the definition of attribute weights, to explicit or implicit interpretation of declarative or other scene properties. Also included in this range is the presentation of example objects in regard to the disambiguation of abstract terms. The

Table 4.2. Preference acquisition

Preference Acquisition Policy	*Applied Methodology*	*Preference Information Acquired*
Transparent (minimal participation)	Incremental learning based on evaluations	Implicit Attribute Importance Subconscious criteria Abstract terms actual interpretation Dynamic Profile
Pre-processing (fixed participation, non-proportional to system use)	Multicriteria attribute weight assignment Example objects	Explicit Attribute Importance Conscious criteria Abstract terms example interpretation Static Profile
Additional feedback during system use (participation directly proportional to system use)	Focused incremental learning Attribute weight adjustment Isolation of interesting properties of inspected scenes	Explicit attribute importance Conscious and subconscious criteria Abstract terms actual interpretation Dynamic Profile

resulting initial profile might be static throughout system use by the specific user and could serve as a reference for a dynamic learning mechanism. In a sense, the comparison with such a static profile could provide an initial estimation of the dynamic mechanism's maturity with respect to the user's preferences. This is particularly useful when considering the incomplete and imbalanced nature of the examples that are produced during regular system use.

Alternatively, the initial profile yielded from the questionnaire could be dynamic itself. As an example, a set of weights for a group of observed scene attributes could be readjusted according to actual evaluations submitted by the user during regular system use. In the case where system functionality is extended with extra requirements for feedback destined for user profile construction, specialized questions may be focused on the currently inspected visualized scenes. Separate grades for the performance of a solution could be requested by the system for a set of observed attributes. Such an approach could be practically combined with the initial profile, gradually improving the latter in a more granular manner than feedback from regular system use comprising only approvals or rejections. Table 4.2 summarizes the alternative degrees of user participation in the profile construction and their corresponding parameters. The integrated intelligent personalization framework presented above is summarized in Figure 4.2. The diagram depicts the subtasks connected with the personalization process as well as their interrelations with the manipulated information.

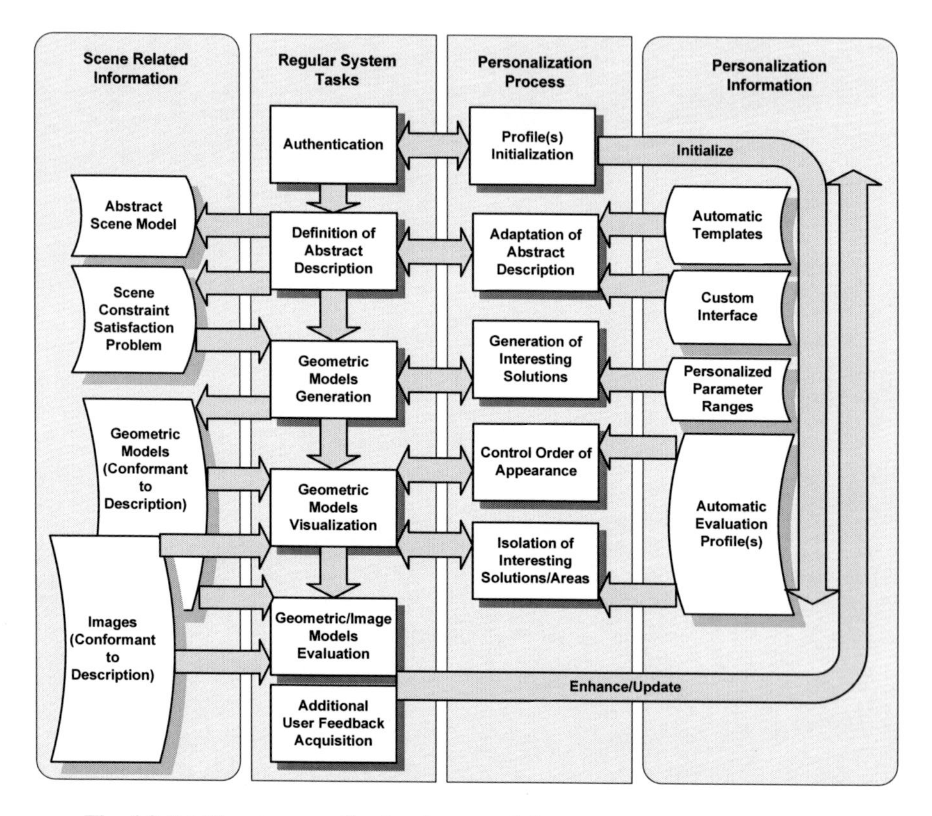

Fig. 4.2. Intelligent personalization framework for a scene modeling environment

4.7 Intelligent User Profile Module Architecture

The framework set in the previous section is employed here through a case study in the context of declarative modeling presented in detail in [1]. The corresponding prototype is responsible for the experimental results presented in a subsequent section. It is interesting to observe in the following to what extent the particularities that arise in a specific environment have guided the selection and adaptation of the aforementioned model in the specific implementation.

4.7.1 Declarative Modeling

A Declarative Modeling environment pertains a number of characteristics which, up to an extent, suggest the translation of the general framework presented above into a concrete module architecture. The major characteristics of such an environment originate from the declarative modeling process itself.

This process commences with the definition of an abstract declarative description of the desired object or environment. This description is subsequently submitted to a solution generation mechanism responsible for the production of valid alternatives, i.e.

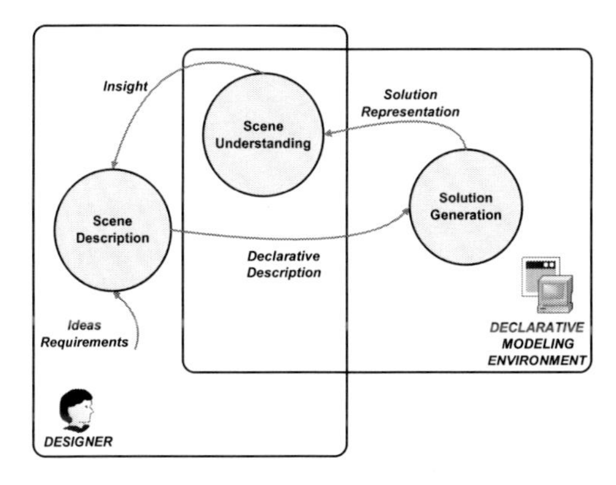

Fig. 4.3. Declarative Modeling Cycle

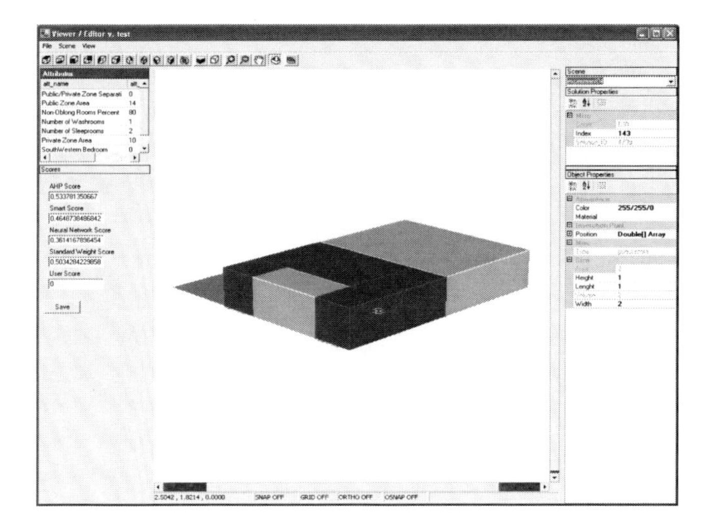

Fig. 4.4. Typical session

geometric representations that all fulfill the input. The solutions are then visualized and presented to the user who, in turn, may select one or more as the answer to the initial description or, alternatively, may refine the initial description thus restarting the declarative modeling cycle. This process is presented in Figure 4.3.

The declarative modeling environment under discussion provides a graphical user interface for the construction of the declarative description of a building as a scene graph. This is subsequently translated into a system of parameter relations (inequalities and equalities) which are submitted to a constraint satisfaction mechanism responsible for producing all alternative valid sets of values for the scene geometric parameters. The latter are eventually visualized and the user is able to inspect and evaluate the solutions, either keeping those of interest or refining the initial description, possibly initiating a new session of system use. A sample session of the system appears in Figure 4.4.

4.7.2 Module Architecture

The architecture of the implemented intelligent user profile module has adopted varied levels of user participation for the acquisition and application of user preferences. In particular, two distinct components operate in parallel towards this direction: a *Machine Learning (ML)* component, solely based on user's input acquired through regular system use, and a *Decision Support (DS)* component, requiring initialization formulated by three alternative multicriteria decision support methodologies. The two components represent the core of the Intelligent User Profile Module. The latter also contains the user profile database, storing personalization information, and the adaptive user interface responsible for the communication with the user and the intelligent presentation of the desired content.

The ML component operates in an entirely transparent manner, being incrementally trained every time a new set of solutions have been inspected and evaluated by the user. This new set serves each time as a new training dataset, aimed to enhance the mechanism's knowledge regarding user's preferences. The user's profile, constructed by the ML component, is practically the incrementally learning mechanism itself. It comprises a committee of neural networks of limited learning ability, i.e. *weak learners*, appropriately combined to form a committee with *strong learner* characteristics. The committee is dynamically enhanced each time a new set of evaluated solutions is available. The degree of committee expansion is directly related with the complexity introduced by the new samples, a complexity expressed through the difficulty (or absence of it) of the existing committee members to correctly classify the new samples. The mechanism is an adaptation of the algorithm presented in [37], modified in order to focus faster on minority samples which, in the current context, represent user approved solutions. The experimental results presented in the following section demonstrate the mechanism's behavior for a group of users under a number of metrics examining different aspects of its performance and reveal its ability not only to enhance its knowledge but also to tackle the evaluation imbalance which is inherent in this application area. Due to its transparent nature, the ML component offers the potential to capture both conscious and subconscious criteria that may have guided the user's evaluations. It is also important to note that the ML component operates under no assumption regarding the similarity or the linear separability of the solutions' classification on behalf of the user, thus representing the widest of the two approaches employed and one of the widest in scene modeling literature [1], [8], [35].

The DS component implements a mechanism requiring user participation which is non-proportional to system use since the input is acquired through a pre-processing step, taking place during the first system use for of each user. The aim of the user preferences acquisition, in this case, is the construction of a weight vector indicating the user's interest for each one of a pre-defined set of observed attributes. Three different multicriteria decision support mechanisms have been adapted to the needs of a

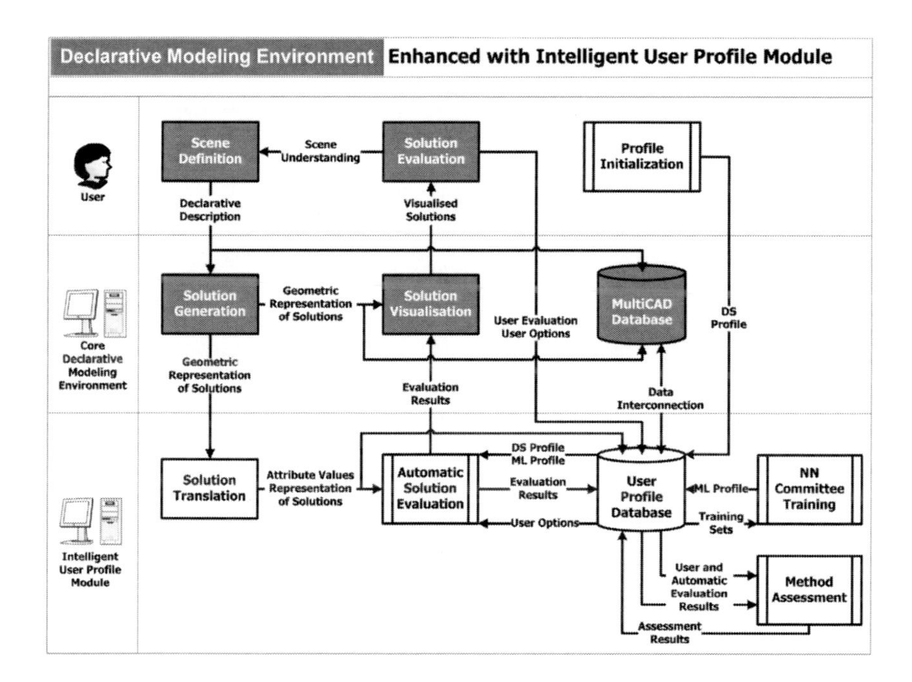

Fig. 4.5. Complete scenario of system use incorporating intelligent user profile module

declarative scene modeling environment [17], [38], [41]. Thus, the user's profile constructed by the DS component is a set of three alternative weight vectors obtained with varying degrees of user's participation. In particular, the first method simply requires the order of importance for the observed attributes, automatically assigning the weights. The second requires the definition of the degree of importance for each attribute when all other attributes are muted. Finally, the third employs detailed pairwise comparison of the attributes, in order to extract the vector weight of their importance. Regardless of the weight assignment method, each solution in this case is automatically evaluated through a weighted sum, thus implying the assumption of linear separability. The DS component represents an efficient alternative since no further training is required during regular system use. As soon as the weight vectors have been constructed, may rely on them for automatic solution evaluation even during the first session.

The application of the user profile constructed by either mechanism takes place during solution visualization, assisting the user at the crucial phase of scene understanding. Solutions are visualized in descending order of calculated preference according to a selected profile, ML or DS. In the latter case the user may choose to adhere to the weight vector of one of the methodologies used or to incorporate more than one during the automatic solution evaluation. Every time a new set of user evaluated solutions are available the performance of the ML component is compared to that of DS component, in order to assess the degree of *maturity* of the former. The components operation and interaction with the environment processes and modules as well as the user-system interaction in the scope of this declarative design system appear in Figure 4.5.

4.8 Experimental Results

In a realistic setup of experiments, users were willing to manually evaluate solutions from one or two descriptions and, henceforth, leave the mechanism take over the evaluation on their behalf. Thus, it is interesting to observe the performance of the module after that point of training. Of course, in order to apply performance metrics, the users had to manually evaluate solutions from additional scenes in order to be able to assess the mechanism's generalization capability.

4.8.1 Performance Indices and Representative Scenes

Thorough evaluation of the mechanism proved to require a collection of metrics capturing different aspects of its performance. In the following, ER is a negative index, representing the overall error (solutions incorrectly classified by the mechanism), HR (Hit Ratio) is a positive index, representing the ability of the mechanism to recall whereas PR (Performance Ratio) is also a positive index, reflecting the mechanism's precision.

In order to cover a wide range of declarative descriptions a set of representative scenes were used in these experiments. Figure 4.6 shows one of these descriptions and one of the solutions corresponding to it.

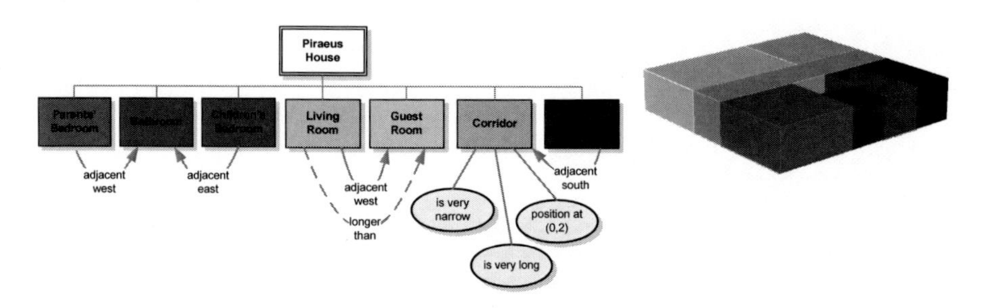

Fig. 4.6. Piraeus house declarative description and a sample solution

In this set of experiments the ML Component was trained every time a declarative description was processed and the user was willing to evaluate the generated solutions. The following tables summarize this evolution, with respect to the ML Component, against solutions corresponding to each scene, before and after the training. The user's profile evolved to a new version every time a set of evaluated solutions was available. In order to gain additional insight, all versions of the user's profile were tested against the evaluated set of solutions for all scenes. Hence, in the following tables, each column represents a new version of the user's profile, after it was also trained with the evaluated solutions of the corresponding scene.

Table 4.3 summarizes the datasets obtained by user evaluation and exposes the inherent imbalance.

Table 4.3. Inspected and approved solutions by two alternative users

Declarative Scene Description	Overall Solution Population	Evaluated Solutions	Approved Solutions (Sophia)	Approval Percentage (Sophia)	Approved Solutions (Katerina)	Approval Percentage (Katerina)
Naxos Habitation	670260	670	41	6.12%	130	19.40%
Rentis Habitation	21184	411	33	8.03%	125	30.41%
Piraeus Habitation	306255	621	47	7.57%	114	18.36%
Kalamaki Habitation	212096	530	17	3.21%	42	7.92%
Patras Habitation	511450	511	52	10.18%	50	9.78%

4.8.2 Experiment Series

Tables 4.4-4.6 exhibit the behavior of the ML Component for one of the users. Each row represents the value of the index under investigation for the solutions of the specific scene. The shaded cells refer to solutions of scenes already used for training. The values in the diagonal refer to the solutions of the last scene used for training. The values under the diagonal refer to solutions from unseen scenes.

Table 4.4 shows an improvement in the generalization ability as new scenes are processed and used for training. It is interesting to observe that previous knowledge may be degraded during training but to an acceptable level, as in the case of Piraeus house solutions, or to the level prior to training, in case that level was already acceptable.

Table 4.5 shows the evolution of the HR that varies significantly both for seen and unseen solutions. This is mainly due to the construction of the algorithm that extends the committee of neural networks with new members based exclusively on the error rate. Moreover, as solutions from new scenes are processed, the approved solutions on behalf of the algorithm become less, thus reducing the HR which signifies the percentage of user approved solutions that are also approved by the mechanism.

Table 4.4. Evolution of ER – 1st Experimental Series (dataset from user Sophia)

Scene \ Version	after Naxos House	after Rentis House	after Pireaus House	after Kalamaki House	after Patras House
Naxos House	14.33%	15.97%	18.36%	18.66%	17.91%
Rentis House	27.74%	20.44%	27.49%	27.25%	27.25%
Pireaus House	67.47%	31.72%	11.76%	15.62%	12.88%
Kalamaki House	25.09%	8.30%	7.36%	5.66%	6.23%
Patras House	10.18%	36.79%	7.63%	9.59%	9.59%
Average Training Error	14.33%	18.20%	19.20%	16.80%	14.77%
Average Generalization Error	32.62%	25.61%	7.50%	9.59%	
Overall Error	28.96%	22.64%	14.52%	15.36%	14.77%

Table 4.5. Evolution of HR – 1st Experimental Series (dataset from user Sophia)

Scene \ Version	after Naxos House	after Rentis House	after Pireaus House	after Kalamaki House	after Patras House
Naxos House	60.77%	52.31%	1.54%	9.23%	10.77%
Rentis House	68.80%	68.80%	8.80%	29.60%	14.40%
Pireaus House	90.35%	94.74%	60.53%	90.35%	52.63%
Kalamaki House	2.38%	57.14%	2.38%	85.71%	4.76%
Patras House	78.00%	100.00%	26.00%	32.00%	62.00%
Average Training HR	60.77%	60.55%	23.62%	53.72%	28.91%
Average Generalization HR	59.88%	83.96%	14.19%	32.00%	-
Overall HR	60.06%	74.60%	19.85%	49.38%	28.91%

Table 4.6. Evolution of PR – 1st Experimental Series (dataset from user Sophia)

Scene \ Version	after Naxos House	after Rentis House	after Pireaus House	after Kalamaki House	after Patras House
Naxos House	82.29%	51.91%	100.00%	75.00%	77.78%
Rentis House	45.74%	68.80%	91.67%	68.52%	85.71%
Pireaus House	41.87%	30.25%	65.09%	47.25%	68.18%
Kalamaki House	3.33%	28.57%	100.00%	58.06%	100.00%
Patras House	26.90%	18.52%	68.42%	21.62%	42.47%
Average Training PR	82.29%	60.35%	85.59%	62.21%	74.83%
Average Generalization PR	29.46%	25.78%	84.21%	21.62%	-
Overall PR	40.03%	39.61%	85.04%	54.09%	74.83%

Table 4.7. Generalization HR (dataset from user Sophia)

Minority Class Generalisation – HR	*After First Scene (Naxos House)*			*After Second Scene (Naxos House + Piraeus House)*		
	Conf. 1	*Conf. 2*	*Conf. 3*	*Conf. 1*	*Conf. 2*	*Conf. 3*
Piraeus Habitation	97.37%	90.35%	100.00%	98.25%	94.74%	100.00%
Kalamaki Habitation	95.24%	2.38%	95.24%	47.62%	57.14%	61.90%
Patras Habitation	30.00%	78.00%	94.00%	100.00%	100.00%	88.00%
Overall	60.52%	66.81%	92.62%	88.35%	88.35%	89.32%

Table 4.6 shows the evolution of the PR, exhibiting the percentage of solutions approved by the mechanism that were also approved by the user. In this case the ML Component seems to considerably benefit from the training, scene after scene, however the generalization ability with respect to the performance ratio varies, obviously depending on the user evaluation which may make the evaluation of solutions corresponding to a scene easier or harder to anticipate.

Table 4.7 summarizes the performance of the ML component with respect to the HR after the first *and* second scene, using three alternative mechanism configurations, where considerable performance improvement is observed. Performance degradation anomalies occur in the Kalamaki House scene having the lowest minority representation as it becomes apparent from Table 4.3.

As a final note, the ML component outperformed the DS component in most experiments, thus revealing the non-linearity of the datasets and the inconsistency exhibited by the users with respect to their DS profile. Subconscious criteria not covered by the observed attributes of the DS component may have also contributed to the ML component's dominance. Further details of the implementation and extensive experimental results are presented in [1], [2].

4.9 Conclusion

In the current chapter we have presented an integrated framework for intelligent personalization in a scene modeling environment. We have discussed the areas which may provide fruitful contribution towards this aim and presented some of the principal mechanisms for this purpose. Subsequently, we have examined in detail the process of intelligent personalization in the context of scene modeling, exploring the interactions among subtasks and corresponding scene representations. Under the light of this analysis, we have proposed a framework comprising a series of proposed interventions for the endowment of a scene modeling environment with intelligent personalization characteristics. An instance of this framework has been employed and implemented in the form of an Intelligent User Profile Module in the context of a Declarative Modeling Environment. The experimental results presented in the last part of the current chapter provide evidence for the efficiency of the implementation as well as of the framework upon which it has been based.

References

1. Bardis, G.: Machine Learning and Decision Support for Declarative Scene Modelling / Apprentissage et aide à la décision pour la modélisation déclarative de scènes (bilingual), Thèse de Doctorat, Université de Limoges, France (2006)
2. Bardis, G., Miaoulis, G., Plemenos, D.: User Profiling from Imbalanced Data in a Declarative Modelling Environment. In: AI Techniques for Computer Graphics, pp. 123–140. Springer, Heidelberg (2008)
3. Borges, J.L.: Labyrinths, Kastaniotis Editions (1992) (in Greek); originally in El jardin de senteros que se bifurcan (1941)
4. Boser, B., Guyon, I., Vapnik, V.: A training algorithm for optimal margin classifiers. In: Fifth Annual Workshop on Computational Learning Theory. ACM Press, Pittsburgh (1992)
5. Brans, J.P., Vincke, P.: A Preference Ranking Organisation Method: The PROMETHEE Method for MCDM. Management Science 31(6), 647–656 (1985)
6. Buckley, J.J.: Fuzzy Hierarchical Analysis. Fuzzy Sets and Systems 17(3), 233–247 (1985)
7. Burbidge, R., Buxton, B.: An Introduction to Support Vector Machines for Data Mining, Keynote. In: 12th Young Operation Research Conference (2001)

8. Champciaux, L.: Classification: A Basis for Understanding Tools in Declarative Modelling. Computer Networks and ISDN Systems 30, 1841–1852 (1998)
9. Chauvat, D.: The VoluFormes Project: An Example of Declarative Modelling with Spatial Control, PhD Thesis, Nantes, France (1994)
10. Dragonas, J.: Collaborative Declarative Modelling / Modelisation Declarative Collaborative (bilingual), Thèse de Doctorat, Université de Limoges, France (2006)
11. Elman, J.L.: Learning and Development in Neural Networks: The Importance of Starting Small. Cognition 48, 71–99 (1993)
12. Expert Choice, Decision Support Software, `http://www.expertchoice.com/about/index.html`
13. Freund, Y., Schapire, R.: A decision-theoretic generalization of on-line learning and an application to boosting. Journal of Computer and System Sciences 55(1), 119–139 (1997)
14. Fribault, P.: Modelisation Declarative d'Espaces Habitable (in French), Thèse de Doctorat, Université de Limoges, France (2003)
15. Goldberg, D.E.: Genetic Algorithms in Search, Optimization, and Machine Learning. Addison-Wesley Publishing Corporation Inc., Reading (1989)
16. Golfinopoulos, V.: Study and Implementation of a Knowledge-based Reverse Engineering System for Declarative Scene Modelling / Étude et réalisation d´un système de rétro-conception basé sur la connaissance pour la modélisation déclarative de scènes (bilingual), Thèse de Doctorat, Université de Limoges, France (2006)
17. Goodwin, P., Wright, G.: Decision Analysis for Management Judgement. Wiley, Chichester (2004)
18. Holland, J.H.: Adaptation in Natural and Artificial Systems. MIT Press, Cambridge (1992)
19. Hornik, K., Stinchcombe, M., White, H.: Multilayer Feed-forward Networks Are Universal Approximators. Neural Networks 2(5), 359–366 (1989)
20. Howard, R.A.: Decision analysis: Applied decision theory. In: Proceedings of the Fourth International Conference on Operational Research, pp. 55–71. Wiley Interscience, Hoboken (1966)
21. Joan-Arinyo, R., Luzon, M.V., Soto, A.: Genetic algorithms for root multi-selection in constructive geometric constraint solving. Computers and Graphics 27, 51–60 (2003)
22. Kechman, V.: Learning and Soft Computing – Support Vector Machines, Neural Networks and Fuzzy Logic Models. MIT Press, Cambridge (2001)
23. Keeny, R.L., Raiffa, H.: Decisions with Multiple Objectives: Preference and Value Trade-offs. J. Wiley & Sons, New York (1976)
24. Konnar, A.: Artificial Intelligence and Soft Computing – Behavioral and Cognitive Modeling of the Human Brain. CRC Press, Boca Raton (2000)
25. Kotsiantis, S., Tzelepis, D., Koumanakos, E., Tampakas, V.: Selective Costing Voting for Bankruptcy Prediction. International Journal of Knowledge-Based & Intelligent Engineering Systems (KES) 11(2), 115–127 (2007)
26. Leshno, M., Lin, Y.V., Pinkus, A., Schocken, S.: Multilayer Feedforward Networks With a Nonpolynomial Activation Function Can Approximate Any Function. Neural Networks 6(6), 861–867 (1993)
27. Lucas, M., Martin, D., Martin, P., Plemenos, D.: The ExploFormes project: Some Steps Towards Declarative Modelling of Forms. In: AFCET-GROPLAN Conference, Strasbourg, France, vol. 67, pp. 35–49. Published in BIGRE
28. Makris, D.: Study and Realisation of a Declarative System for Modelling and Generation of Style with Genetic Algorithms: Application in Architectural Design / Etude et réalisation d'un système déclaratif de modélisation et de génération de styles par algorithmes génétiques (bilingual), Thèse de Doctorat, Université de Limoges, France (2005)

29. Martin, D., Martin, P.: PolyFormes: Software for the Declarative Modelling of Polyhedra. The Visual Computer 15, 55–76 (1999)
30. Miaoulis, G.: Contribution à l'étude des Systèmes d'Information Multimédia et Intelligent dédiés à la Conception Déclarative Assistée par l'Ordinateur – Le projet MultiCAD, Thèse de Doctorat, Université de Limoges, France (2002)
31. Miaoulis, G., Plemenos, D., Skourlas, C.: MultiCAD Database: Toward a unified data and knowledge representation for database scene modelling, 3rd 3IA International Conference on Computer Graphics and Artificial Intelligence, Limoges, France (2000)
32. Mitchell, M.: An introduction to Genetic Algorithms. MIT Press, Cambridge (1998)
33. Multiple Criteria Decision Analysis – State of the Art Surveys. Springer, Heidelberg (2005)
34. Plemenos, D.: Declarative modelling by hierarchical decomposition. The actual state of the MultiFormes project, Communication. In: International Conference GraphiCon 1995, St Pe-tersburg, Russia (1995)
35. Plemenos, D., Miaoulis, G., Vassilas, N.: Machine learning for a General Purpose Declarative Scene Modeller. In: International Conference GraphiCon 2002, Nizhny Novgorod (2002)
36. Plemenos, D., Sokolov, D.: Intelligent Scene Display and Exploration. STAR Report. In: International Conference GraphiCon 2006, Novosibirsk, Russia (2006)
37. Polikar, R., Byorick, J., Krause, S., Marino, A., Moreton, M.: Learn++: A Classifier Independent Incremental Learning Algorithm. In: Int. Joint Conf. Neural Networks, pp. 1742–1747 (2002)
38. Roberts, R., Goodwin, P.: Weight Approximations in Multi-attribute Decision Models. Journal of Multicriteria Decision Analysis 11, 291–303 (2002)
39. Roy, B.: Classement et choix en présence de points de vue multiples (la méthode ELECTRE). RIRO 8, 57–75 (1968)
40. Russel, S., Norwig, P.: Artificial Intelligence – A Modern Approach, 2nd edn. Prentice-Hall, Englewood Cliffs (2002)
41. Saaty, T.L.: The Analytic Hierarchy Process. MacGraw-Hill, New York (1980)
42. Schapire, R.E.: The strength of weak learnability. Machine Learning 5(2), 197–227 (1990)
43. Turing, A.M.: Computing Machinery and Intelligence. From Mind LIX (2236), 433–460 (1950)
44. Vapnik, V.N.: Statistical Learning Theory. Wiley, Chichester (1998)
45. Vincke, P.: Multicriteria Decision-aid. Wiley, Chichester (1992)
46. Visa, S., Ralescu, A.: Issues in Mining Imbalanced Data Sets - A Review Paper, Sixteenth Midwest AI and Cognitive Science Conference, pp. 67–73, Dayton, April 16-17 (2005)
47. Weiss, G.M., Provost, F.: Learning When Training Data are Costly: The Effect of Class Distribution on Tree Induction. Journal of Artificial Intelligence Research 19, 315–354 (2003)
48. Witten, I.H., Frank, E.: Data Mining – Practical Machine Learning Tools and Techniques, 2nd edn. Elsevier, Amsterdam (2005)
49. Zhang, J., Mani, I.: k-nn Approach to Unbalanced Data Distributions: A Case Study Involving Information Extraction. In: Proceedings of the ICML-2003 Workshop: Learning with Imbalanced Data Sets II, pp. 42–48 (2003)

5

Web-Based Collaborative System for Scene Modelling

John Dragonas and Nikolaos Doulamis

Department of Informatics, Technological Educational Institute of Athens
Ag. Spyridonos St., 122 10 Egaleo, Greece
idrag@teiath.gr, ndoulam@cs.tua.gr

Abstract. Contemporary scene design problems are inherently complex and require multiple designers to work collaboratively. This chapter is concerned with the contribution to the development of collaborative systems for declarative modelling by hierarchical decomposition in scene modelling. Furthermore, it presents the development and evaluation of an effective collaborative prototype environment which facilitates designers to exchange and share information through the Internet during the design projects. In particular, the implementation of the web-based system concerns the *Collaborative Declarative Modelling System* framework as an extension of MultiCAD software architecture. This framework is based on a specific module that was designed in order to exploit the declarative model content in the implementation of the collaborative mechanism. We present the appropriate Web-based collaborative prototype workspace developed as a MultiCAD Client. We also describe team profile estimation mechanisms able to update the proposed collaborative prototype response with respect to the current team's information needs. We investigate two different approaches. The first estimates the designers' profiles by considering each designer to be independent from each other. On the contrary, the second method exploits the dependencies of the designers involved in a design. As far as the first approach is concerned, a recursive algorithm is presented based on a neural network architecture. Regarding the second approach, we implement two different algorithms; a) the preference consensus scheme in which the degree of relevance for a solution is defined by the preferences of all the involved designers and b) a spectral clustering scheme in which designer classes are constructed from the designers preferences in a way that inter class information is minimized, while the intra class flow is maximized. Finally an example of use of the developed prototype system as well as its formal operation is presented.

Keywords: Collaborative Scene Modelling, Declarative Modelling, Collaborative Design, Data Sharing, Synchronising Group Operations.

5.1 Introduction

In the past years, computer science specialists have focused on systems known as "collaborative". Collaborative architectures yield a more productive cooperation between workers, students, designers and other groups. Thanks to recent technological evolution, like for example the Internet, Collaborative Systems (CSs) become much easier, efficient and effective, increasing the mutual cooperation between individuals or groups of common objectives. The need for collaboration appears when individuals fail to carry out a task on their own because of their limited abilities (lack of knowledge or

G. Miaoulis and D. Plemenos (Eds.): Intel. Scene Mod. Information Systems, SCI 181, pp. 121–151.
springerlink.com

strength), or when collaboration can help them to carry out this task faster and more efficiently. It has been experimentally noticed that any task could be achieved much faster and with better results in a collaborative or on a network environment, when the members of a group execute the work in parallel. Moreover, information access is faster when information is shared in a common database under an architectural framework that enables persons, taking part in a project, to collaborate with each other. Furthermore, CSs yield a reduction of the cost and consequently an increased profitability, as communication and exchanges are electronically managed. If several different individuals from the same computer environment were asked what CSs are, we would probably get a lot of different answers. Some would say that collaboration or communication is the e-mail. Others would refer to teleconferencing and the World Wide Web. One could even hear references to modern ways of discussion and the chat. Generally, it is difficult to define this concept, because of the great number of technologies available today. Actually, each answer mentioned above is correct. The CSs result partly from the use of different technologies in the same environment, and their purpose is to facilitate and promote the distribution of information and the management of work and communication. However this technology "unification" represents only one part of the collaboration as we define it. Another part is *synchronization*. Everybody knows about the applications on which people work altogether simultaneously with others and then each person works and gives ideas in turn. But now, new technologies offer another way of collaboration, the so called *non-synchronized collaboration* in which the participants do not need to be together and they can collaborate anytime they wish. Examples of non-synchronized CSs are the e-mails, the Internet, the intranets and shared databases.

In [13] it is presented that conception can be considered as a series of separate activities, and that the level of the specialization is determined by the way the task would be carried out and by the time necessary to achieve it. Usually, a design consists of a series of very particular steps. At the beginning, collaborators work together but, then, they have to separate and work on their own. They behave like distinct specialists who always look at things from their own perspective. Their specialization can change during the process because their understanding is complementary and it is through participating to this process that they learn.

This analysis is in accordance with a model of "Collaborative Design (CD)" which consists of short-lasting parallel actions that are framed by a concerted action of discussion and evaluation. Thus, the design act is distinct, isolated, parallel and after all not completely connected to the others. In that model, the work is naturally cooperative and the cooperative aspect concerns the estimation and the evaluation, which are essential activities in the Computer Aided CD. This point of view is also supported in the categorized classification of the CD set out in the experiment of a team [19]. More specifically, one can distinguish three categories. First, the shared collaboration, in which participants works together. Second, the exclusive collaboration, in which participants works on different parts of a problem by discussing and consulting each other regularly. Third, the "dictatorial" collaboration, in which participants decides who is going to be their leader and be in charge of the procedure.

As a matter of fact, it is in the exclusive collaboration that the team has observed the most productive results. As far as shared collaboration is concerned, it has had no result after a really intense exchange between the participants. Finally, the "dictatorial" collaboration comes to a solution only when the leader makes the final decision. Therefore, exclusive design, in comparison with a design achieved in a simple collaboration, requires a higher sense of cooperation in order to reach a more creative result. Thus exclusive design is much more demanding, more difficult for a group to set up and to continue than just trying to reach one goal. We should mention the fact that collaboration does not include re-examination of the decision by the members of the team, separately or after agreement. However, when we look a little deeper in the collaboration process, another problem appears; the concessions that the participants have to make sometimes on their demands, yielding a badly organized system that does not entirely satisfy the participants' information needs and preferences.

A solution for this problem is proposed in [8] that is concerned a compromised framework in which designers, who are going to collaborate with each others, will negotiate to reach to common conclusions, which are sufficient, but not optimal. Conclusion to a non optimal solution does not mean that this solution should be rejected, since it represents the average satisfaction for all participants. Probably, one of the most popular collaborative systems is in the era of building design, which is inherently multi-physical and multi-disciplinary; no single person is able to perform a full development process for a building feature or system. Recently, in the research community the notion of "Collaborative Product Development" (CPD) systems has presented. These systems are defined as "an Internet based computational architecture that supports the sharing and transferring of knowledge and information of the product life cycle amongst geographically distributed companies to aid taking right engineering decisions in a collaborative environment" [35]. During the last years, many technologies have been presented in the literature in the field of CPD. Their focus has been on sharing product data and providing collaborative tools to bring multidisciplinary teams together [7].

5.1.1 Research Scope

In this chapter, we present the following two research issues. The first concerns a WWW-based platform for collaborative design of a particular area of scene modelling that of early phase architectural design. The proposed "Collaborative Declarative Modelling System" (CDMS) is based on client-server architecture, and allows multi-designers/designers in different locations to participate in the same design process under a declarative design framework. The system was designed and implemented in order to provide designers a working environment for the development of large architectural project(s), in a synchronous or asynchronous way, even if the participated designers are geographically located in distant workplaces. In the present project, the CDMS is analysed and the collaborative mechanisms that have been developed are clarified. These mechanisms are designed in a way that exploits the concepts of collaborative/declarative design. In addition, we present a case study which reveals the operational and functional capabilities of the system as well as implementation issues and benefits arising from the usage of the collaborative framework.

The second research issue is the intuitive manner that the designers describe the scenes. Scene description is often imprecise for mainly two reasons. The first arises from the fact that the designers do not know exactly the characteristics of the scene before the design. The second reason refers to the inaccurate description of the design-requirements. For example the statement, "put object A on the right of the object B" may result in different interpretations as far as the exact position of the objects in the scene is concerned [32]. To address the humans' subjective description of a scene, personalization mechanisms are incorporated in the proposed architecture. These tools filter the generated solutions according to the current designer's information needs. In this way, we can derive more intelligent retrieval mechanisms. Two different scenarios are presented regarding designer profile estimation. The first assumes that the involved designers are independent from each other. On the contrary, the second approach exploits the dependencies among the designers. As far as the first approach is concerned, two versions are discussed. The first implements a neural network model for estimating the degree of relevance between a stored object and the scene description. The second method extends the first algorithm by introducing a recursive schema for the neural network training. Regarding the second approach, we have implemented a preference consensus model and a spectral clustering algorithm. The preference consensus evaluates the solutions according to the degree of relevance of all designers. Instead, the second schema classifies the designers in clusters according to their preferences. Experimental results have been presented which indicates the performance of the proposed designer profiling algorithms.

This chapter is organized as follows: in section 2, we present previous related work. In section 3 we present and analyse the implementation framework of the Collaborative Declarative Modelling System, and we describe the components of the collaborative system. For the aims of this research, we employ the above system during the early phase of the architectural design process. In section 4, we apply a case study in order to prove the functionality of the developed system. The designer profile algorithms (single and collaborative approach) are described in section 5, while conclusions are drawn in section 6.

5.2 Related Work

5.2.1 Collaborative Design

The recent advances in broadband technologies and networking stimulate the need for network-centric Computer-Aided Design/Manufacturing (CAD/CAM) environments came [37]. Integral parts of these environments are collaborative modelling systems, [41], [17], [43], where designers can use interactive modelling sessions. Collaborative geometric modelling systems are required for designers located in multiple geographic areas. This method of concurrent engineering may avoid several iterative steps, which would otherwise be required by the design process. The collaborative CAD systems – depending on the collaborative level that they support – could be categorised in four categories [14] depending on the collaborative functionalities that they support; from the basic collaborative environment to a full functional one.

5.2.1.1 Collaborative Systems

Several collaborative systems have been developed in the literature which they exploit the internet technology for designing a product. The review of the integral arts, which a collaborative environment consists of, is very important for understanding the technological aspects required for an efficient implementation of a collaborative environment. The main parts of a collaborative system are the following:

- Information Systems Architecture. A structured and transparent framework is generally adopted for the architectural design of a collaborative environment [25].
- Communication Tools. They allow visual and audio communication between the geographically dispersed group members.
- Virtual tools for group management. They coordinate the dispersed group members.
- Geometrical Representations. A software application that facilitates the visualization of a product design among all the geographically dispersed group members.
- Integration Software. Application interfaces for import and export files from commercial CAD/CAM/CAE systems.
- Knowledge Representation Methods. Generic rules that are stored in an information database and describes conceptual structures of a design.
- Project Management Tools. They coordinate the activities of product development.

As far as the CAD architectures are concerned, there exist two principal categories concerning the collaborative systems. First, CAD applications, which use distributed workspace, (for example: 'ARCADE'). Second, CAD applications that are based on the Internet (for example, Buzzsaw Standard™). The recent collaborative architecture consists of a combination of distributed workspaces, which are accessible thanks to recent progress in networking (e.g., Virtual Private Networks). There are many research projects on the era of Collaborative design. We briefly describe some of them.

The 'CSCW-FeatureM' is a CAD system [39]. In this system the clients are responsible for replication and data processing. The modifications are done consecutively after prior agreement. The consistency check is done at the end of the session by comparing the histories. The whole model is shared. There are no server but 'FeatureL' clients (ACIS™). The 'CollIDE' is a synchronous shared workspace [27]. It provides replication on both clients and server, whereas data processing is done on the client. The level of sharing relates to part of an assembly. The server manages the sessions and is used as a database. The clients have a CAD application (Alias Studio™) for data publishing as well as an additional system to visualize shared geometry. The collaborative component is Group Kit [36]. The 'TOBACO' [9] is a CAD system that supports replication and data processing on clients. The whole model is shared. The server deals with management and history of the sessions. The clients have a CAD application (ACIS™/AUTOCAD™). The 'ARCADE' environment [38] provides replication and data processing on clients. Only one participant can modify the objects at a time. Only the objects are shared. The server directs the management of the sessions and the database. The clients have a CAD application (ACIS™). The project 'Collaborative Solid Modelling' [6], has a client/server architecture. In each client, a client manager controls the sending of data to the server and receives data from the server. At the server, a session manager controls one or more groups and the design geometry. Each group has one or more clients. All the clients share the same objects within a group. The project 'NetFeature' is a CAD system [18]. Both clients and servers are responsible for replication and data processing. However, it does not manage access conflicts. The level of sharing relates to the whole model. The servers deal the management of a session and the design geometry based on the commercial geometrical modeller of ACIS™. The database is located on another server. The clients manage the Web navigator (visualisation is based on Java3D™). The project 'Web Based Collaborative CAAD' is a CAD system for Architectural-Design [1]. The architecture is based on the replication of the data on the clients and a server. However, the processing is carried out on one or more servers. Each designer works on a different part of the model. The level of sharing relates to part of a model. The server manages the sessions, geometrical modellers and a database. The clients have an interface and a Web navigator for model visualization (Java™/VRML). The component of collaboration is Shastra [1]. The 'WebSPIFF' is a feature-based modeller who allows simultaneous feature creation, removal and modification by many designers [5]. The server is responsible for replication and data processing. The operations are carried out with 'Co-Create'. The level of sharing relates to the data. The server manages sessions, geometry (ACIS™) and Web services. The clients have Web navigator for visualization (VRML-Java3D™, or fixed images). The project 'Syco3D' is a CAD system [28]. The clients are responsible for replication and data processing. The level of sharing relates to part of an assembly. The servers manage the sessions. The clients have a CAD application for data publishing and an additional window to visualize shared data. The component of collaboration is 'GroupKit' [36]. The project 'CyberCAD' [40], is a Web-interactive, designer-friendly software. It allows the development of 3D models and supports synchronous collaboration between geographically dispersed designers. 'CyberCAD' became a simple, robust, portable, and multithreaded, with independent distributed and dynamic platforms.

5.2.2 Declarative Design

The declarative scene modelling is based on declarative conception cycle, which consists of three sequential functional phases [30]:

1. The *Scene Description* phase, which permits the description of properties of the scene to be created.
2. The *Scene Generation* phase, which permits the production of scenes verifying the description given by the designer.
3. The *Scene Understanding* phase, which allows the designer to confirm that the solution scenes satisfy the required properties.

Figure 5.1. represents the basic declarative conception cycle. In the classic collaborative CAD systems, data sharing is implemented in the geometrical modelling phase while in collaborative declarative systems it is implemented in the scene description phase respectively [33].

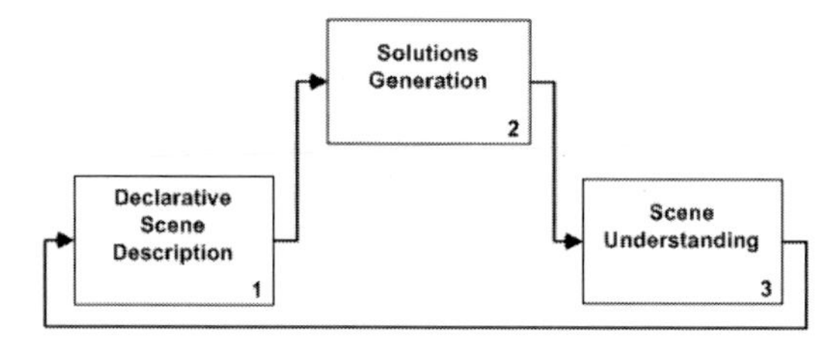

Fig. 5.1. The declarative design cycle

5.2.3 Overview of MultiCAD Architecture

MultiCAD is an intelligent multimedia information system that is based on declarative modelling [23], developed by the Laboratory Méthodes et Structures Informatiques of the University of Limoges along with the team of Information Systems & Applications in the Department of Informatics of the Technological Educational Institute of Athens. It was launched as a research project for developing a multimedia information management system, supporting the generation of geometric models from a description of a scene based on a declarative modelling hierarchical decomposition (DMHD) schema [31]. This software framework is a platform for developing multimedia information management systems supporting all phases of the declarative process, [22]. Furthermore MultiCAD architecture is able to support a collaborative design environment [14]. There exist different system prototypes based on MultiCAD architecture [3], [12], [15] [24], [32], [34], [42]. Figure 5.2. presents a typical MultiCAD session.

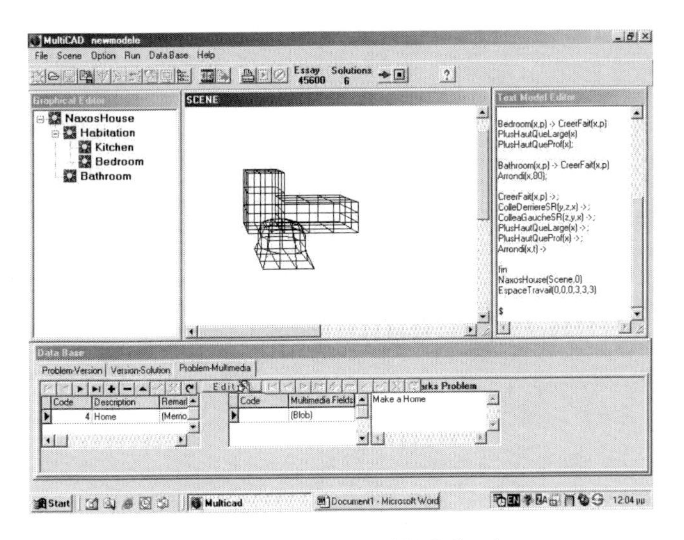

Fig. 5.2. A Typical MultiCAD Session

In order to model a scene in MultiCAD, the designer can describe, in an intuitive manner, the scene's composing objects, their properties and relations. There exist two kinds of representation of the scene:

- text-formed representation, expressed in a formal Prolog-like language [31], representing the "Internal Declarative Representation" (IDR) or internal model of the scene,
- relational-object representation [22] in which a storage system is organized in four logical databases: The "Scene database" is supporting information describing the scene models (IDR models, relations between the composing objects etc.), the "Project database" is manipulating data concerning the planning, the finance and other specially oriented data for each project, the "Multimedia library" contains all types of documents related to a project (geometrical models and multimedia information as pictures and videos) and the "Knowledge database" which contains the necessary information about entities' types, their properties and relations.

The latest version of the conceptual model of MultiCAD architecture [22], is based on the Extended Entity-Relationship model. Such model extends the Entity-Relationship model in order to include notions as aggregation, inheritance etc. The scene's description is stored in the relational-object representation as an assembly of Objects having Properties and being related to each other through Relations. Furthermore, the Knowledge database contains a series of Types of Objects, Types of Properties and Types of Relations constituting the description of Knowledge in MultiCAD. The elements of the Scene meta-model are instantiated in the Knowledge database.

5.2.4 DKABM Framework

In order to establish a collaborative design environment it is necessary to adopt a representation that follows declarative modelling principles and architectural principles, building modelling and architectural knowledge. One such approach is introduced in [20] and is called *Declarative Knowledge Framework for Architecture-oriented Building Modelling* (DKABM). The DKABM could cover all information required to consistently and completely describe any building scene from an architectural point of view [21]. The adoption of alternative DKABMs in MultiCAD architecture incorporates the conceptual design of scenes with domain-specific knowledge 33. The description of a building is organized in categories of objects with same properties and relations. Information is organized hierarchically forming a knowledge decomposition tree. All tree nodes are entities, (i.e., categories of objects), having properties, Intra and Inter-relations, such as i) descriptive and structural properties, ii) functions of an entity, iii) declarative properties, iv) inner-relations, v) generalized relations, vi) aggregation relations and finally vii) further associations. The DKABM framework facilitates the identification of semantic relationships needed to define architecture-oriented building models. In addition, architectural constraints have been determined in order to validate the coherency and the semantics of the represented buildings [20], [21].

5.2.5 Declarative Design Representations

A designer can model a scene by describing the objects that a scene is composed to, their properties and relations. A scene can be represented in a text-formed representation, expressed in a formal language [30], as *Internal Declarative Representation* (IDR) and as a *Scene Graph*. According to MultiCAD architecture [22] the scenes models (IDR models, relations between the composing objects etc.) are stored in a Scene database. The Scene database is configured following the Scene Conceptual Modelling Framework (CMF) [22], the entities of which have been described in previous section. The *Scene Database* [23] is configured following the Scene CMF where the description of a scene contains entities, such as. The **Objects** defined by their properties, simple or generic ones, as well as group of simple objects with properties in common. Three types of **Relations** between objects: **meronymic** (is-part-of, is-included-in), **spatial organisation** (near-by, on-left) and **inter-scene** (higher-that-large, higher-than-deep) relations. The **Properties**, which they describe objects. Finally, a *Scene Graph* represents a scene by connecting all the involved entities. Objects, Relations of objects as well as Properties of objects are combined in a graph to represent the scene.

5.2.6 Collaborative Declarative Modelling System

The Collaborative Declarative Modelling System (CDMS) has been presented in detail in [14] and comprises the following main parts, (Figure 5.3.):

1. Declarative Collaborative and Workspace analysis module (DCM) which includes collaborative modelling primitives and declarative models analysis,
2. MultiCAD Client Browser (Collaborative Workspace),
3. Server facilities (Communication layer, Service calls, session manager, Communication facilities),
4. Distributed MultiCAD data & knowledge resources (MultiCAD Databases).

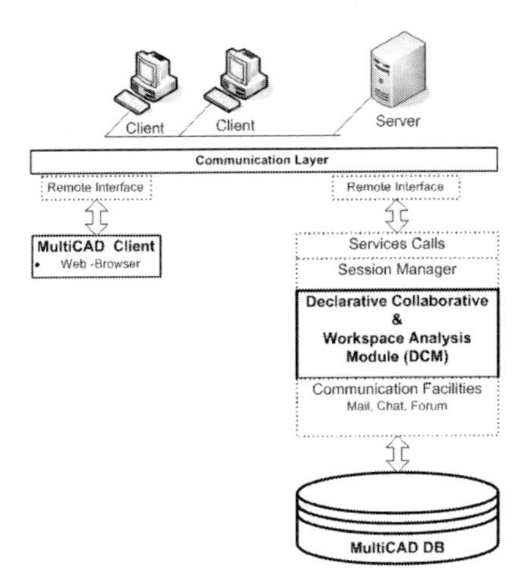

Fig. 5.3. The Collaborative Declarative Modelling System

5.3 Web-Based CDMS Framework

The main objective of this section is the description of a web-based framework for CDMS [11]. This framework operates in the HTTP environment as communication layer. The systems facilities (services calls, session managers etc) are assured by the operating system. The distributed data and knowledge resources enclosed in the term of MultiCAD database are concretized for the needs of this work under a set of related tables. The system development is based on open source software technology. In particular, the system software scripting is made of PHP pages, while the RDBMS is MySQL. The system is installed on Windows 2000 Advanced Server platform. Its flexibility made it easily transferable into a Linux, or any type of Unix-based web server, without the need for further conversion or addition, but only these concerning its visual appearance. The demands from the part of the client are minimum and the only need for an unobstructed functionality of the client is a current version of a web browser. The functionality of the system is based on the overlapping of the database by PHP pages that implement the logic structure and the algorithmic process of the collaborative mechanism. These pages built a designer interface with specific abilities

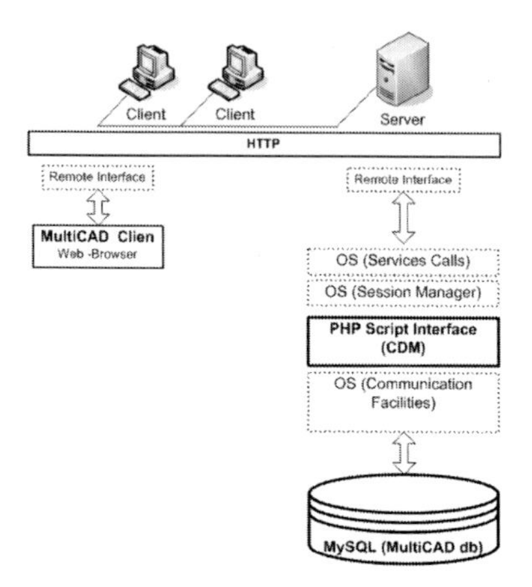

Fig. 5.4. The Web-based CDMS in MultiCAD environment

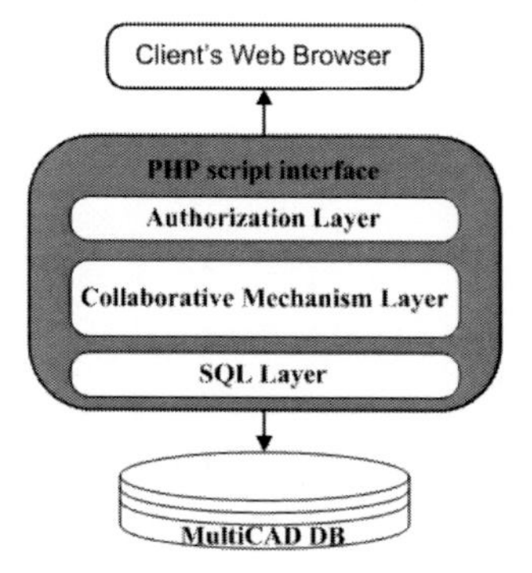

Fig. 5.5. How System framework restricts client access to the collaborative database. System Components of web-based CDMS framework.

and restrictions. The designer actions are restricted to actions that are allowed from the PHP pages. The Information regarding designers' connections and designers' profile is stored in a database, which is controlled by the system administrator. Figure 5.5. presents the architectural components of the Web-based CDMS implementation in MultiCAD environment.

The designer accessibility to the collaborative database is restricted by:

1. Authentication process
2. Collaborative mechanism
3. SQL syntax

Any activity of a designer to the collaborative database must comply with the above three layers of restrictions otherwise it is prohibited (Figure 5.5.).

5.3.1 Declarative Collaborative Module

The Collaborative System consists of three layers of restrictions, which support the management of the collaborative database [11]. At any time, any designer who tries to

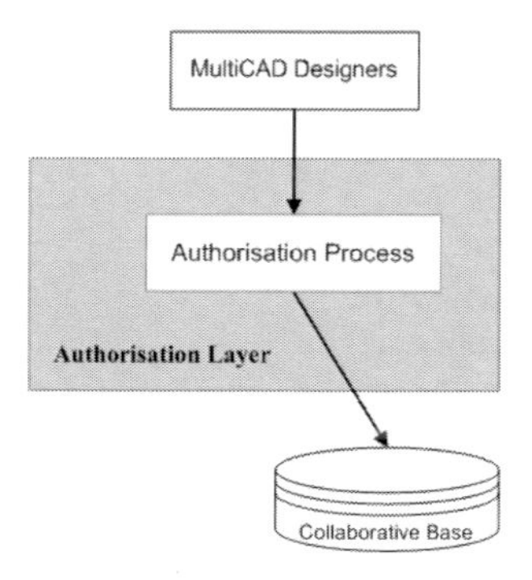

Fig. 5.6. The authorization layer

access the collaborative database is forced to encounter all these layers and check against each one of them independently. In case of successful pass through each one of them the designer is allowed to alter the collaborative database.

Authorization Layer. The collaborative database contains certain information for each designer such as: name, location, access rights, designer name, password etc. PHP pages take on the responsibility of the authentication process for a designer by comparing the designer name-password and the relative information that the collaborative database retains (Figure 5.6.). The collaborative database retains designers' information in a specified table with the following schema.

```
id int(6)
login_name varchar(8)
pasword encrypted
name varchar(30)
surname varchar(30)
location varchar(50)
last_login varchar(30)
access varchar(10)
PRIMARY KEY (`Id`)
```

In case of successful authentication of a designer, his/her data (designer name-password-access) are passed to the next layer (Collaborative Mechanism Layer). Based on the "access" field value certain function buttons may or may not be visible and certain rights to projects are loaded. In this phase of development the possible values are "Admin", "Viewer" or the name of the project in which the designer belongs. Each successful login is recorded for administration purposes in a table with a schema as follows:

```
id bigint(20)
login-datetime varchar(40)
logout-datetime varchar(40)
userer varchar(8)
ip varchar(15)
PRIMARY KEY (`id`)
```

Collaborative Mechanism Layer. The collaborative database retains the architectural projects as scenes, and every scene description (blueprint) as a DKABM. Each DKABM is saved in a table. This table's schema has the following form.

```
record_id bigint(20)
father_id smallint(6)
child_id smallint(6)
PRIMARY KEY (`record_id`)
```

For every node there are as many records as its children. This table can be constructed using a specific function button which is available only for the system Administrators. This way the environment can contain multiple DKABMs. However, a project is related only to one DKABM. Another table retains the actual objects that take part in a scene at any time with the following schema.

```
Object_ID int(11)
Obj_Name varchar(50)
Obj_Descr varchar(50)
Object_Type_ID smallint(6)
Obj_Type_House varchar(50)
PRIMARY KEY (`Object_ID`)
```

The 'Object_Type_ID' record is related to the 'father_id' and 'child_id' records of the DKABM table. An authenticated designer can load up a PHP page where he/she can view current project(s). This page shows the scene decomposition to a conceptual tree based to a DKABM. This tree contains all objects of the scene.

A node in the conceptual tree represents every object and any decomposition relation to one or more children is represented by a connection. This PHP page is the interface that the designer uses in order to select a scene object to operate on. For the automatic construction of the conceptual tree a specific PHP recursive function is used (Figure 5.7.).

```
function makethetree ($node_pos, $objects_array) {
  $children_array = read_the_children($node_pos);
    if ($children_array == 0) {
        add_terminal_node($node_pos);
        return;
    } else {
        foreach ($children_array) {
            $father = $node_pos;
            $child = next_child($children_array);
            add_a_father($father,$child);
            makethetree ($child, $objects_array);
        }
    }
}
```

Fig. 5.7. Conceptual tree construction algorithm

For the proper operation of this function it is necessary to have a matrix (linear array) containing all the objects already created, which is given as the second parameter $objects_array. The main functionality is based on a JavaScript function which creates the optical entity of the conceptual tree. The above function among other secondary functions is provided by the PHP script interface and is implemented in the server-side. A designer can apply to an object the following operations. "View" (Unrestricted operation), and "Modify" (Restricted operation).

The operation "View" is provided unrestricted to any authorized designer at any time. The operation "Modify" is a restricted operation by the collaborative mechanism. A designer in order to apply a "Modify" operation to an object-(node) must register himself/herself for it. The restriction mechanism allows or denies this registration depending on the state of this object-(node). Also a registered designer is limited only to a selected object, and he/she cannot register for the same or another object twice.

The object can have one of the following states. "Registered" to another designer. "Closed" as a "Child" node of a tree registered to another designer. "Closed" as a "parent" node of a tree that contains another node registered to another designer. "Open", for possible registration. In the conceptual database, there are two tables containing this kind of information. The table 'analysis' contains information about which designer handles which object, and has the following form

```
id mediumint(9)
user varchar(8)
state varchar(20)
object int(11)
PRIMARY KEY (`id`)
```

The table 'closed_objects' contains all the closed objects, with the schema,

```
id mediumint(9)
object int(11)
cls_state varchar(10)
PRIMARY KEY (`id`)
```

As Table 5.1 shows, the registration to an object (node) is prohibited on certain cases except the case of either an open object (node), or a closed parent state when the operation is allowed.

Table 5.1. The relationship between the state of a node and the Registration operation

Node state:	Registration:
Registered	Not Allowed
Closed	Not Allowed
Closed parent	Allowed
Open	Allowed

The general condition is the following: when a designer registers himself to an object-(node), he/she defines the state of that object (node) as "Registered", and additionally the objects-(nodes) of the sub-tree that might contain this object inherit as the "Closed" state. In particular, "Registered" object any contained object in its sub-trees is in the state "Closed parent" (Figures 5.8., 5.9., 5.10.).

All these different states of the objects are saved in "analysis" table, along with information about the designer who affects the objects.

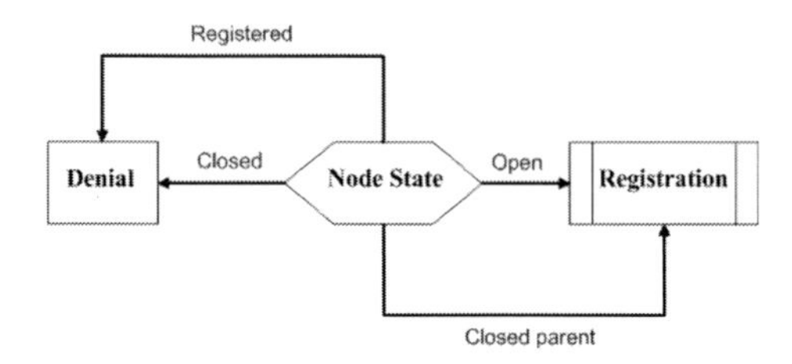

Fig. 5.8. Node registration algorithm

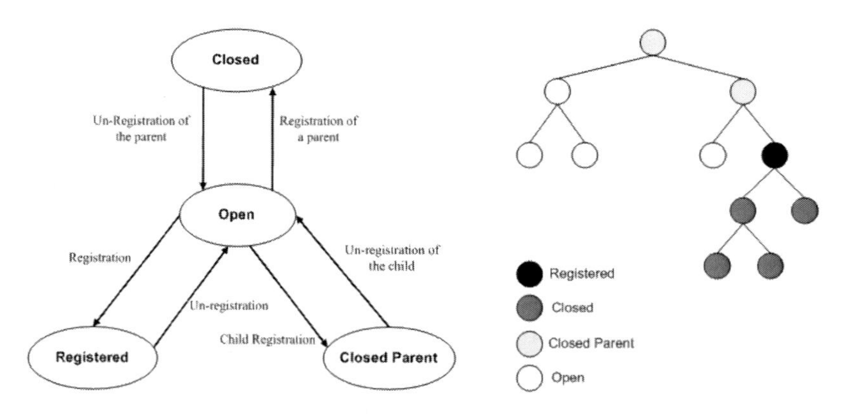

Fig. 5.9. Different states of an object

Fig. 5.10. A typical situation of a tree with one registered node

Since a designer is registered to an object-(node), he/she can apply the "Modify" operation. The "Modify" operation is decomposed in the following sub-operations. Modify object's Details, Modify object's Properties, Modify object's Relations, Insert another object as child to the registered object, and Delete one of the direct children objects of the current registered object.

Each of the above sub-operations is available for editing the objects-(nodes) of the sub-tree of the registered object-(node) as well. The only exception is that the sub-operations 4 and 5 can be applied only to registered parent objects-(nodes).

Table 5.2. The relationship between the state of a node and the Modify-Insert-Delete operation (for another designer)

	Registered by another designer	Closed by another designer	Closed as parent by another designer
Modify Details	N/A	N/A	N/A
Modify Properties	N/A	N/A	Allowed
Modify Relations	N/A	N/A	Allowed
Insert another object	N/A	N/A	Allowed
Delete this object	N/A	N/A	N/A

Table 5.3. The relationship between the state of a node and the Modify-Insert-Delete operation (for the same designer)

	Registered to this designer	Closed to this designer
Modify Details	Allowed	Allowed
Modify Properties	Allowed	Allowed
Modify Relations	Allowed	Allowed
Insert another object	Allowed	N/A
Delete From this object	Allowed	N/A

Each sub-operation might have certain restrictions dependent of the state of the object that is going to be applied to. Tables 5.2, 5.3 are shows particular combinations of operations among/about objects.

The "Modify" sub-operation edits the objects Properties**,** including these operations. "Add" a new property, "Delete" a property, and "Update" an existing property. The "Modify" sub-operation edits the objects Relations including these operations. "Add" a new relation, "Delete" a relation, and "Update" an existent relation. The state of an object-(node) at any time, for a specific designer, also affects how its relations with other objects (nodes) can be considered. As a mater of fact, the relations of a registered parent object-(node) with its child object(s) within its tree, are considered as "Local", and the relations with all other objects are considered as "Global". The designer can use sub-operation "Modify" to a relation only if this relation is "Local". Figure 5.11. shows an example of the relations of a registered parent-object, and its sub-tree with its child-objects.

All information that concerns a sub-operation (operation, object, and other data) is sent to the SQL layer for further execution (Figure 5.12.). It is necessary to mention here that critical information (properties, details, etc.) about objects that take part

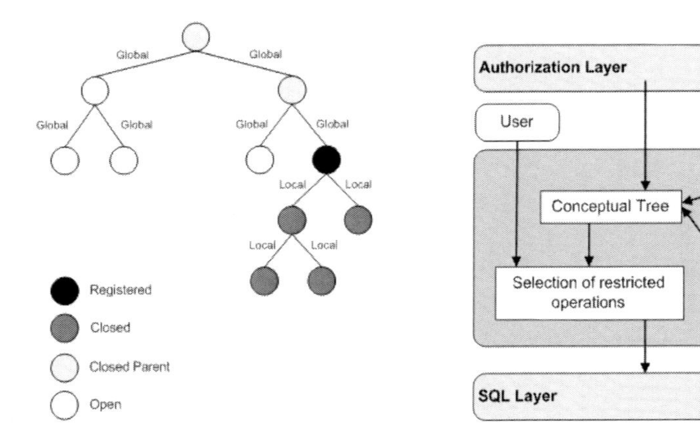

Fig. 5.11. The relations for a registered object

Fig. 5.12. The collaborative mechanism layer

in a particular project is contained in specific tables called in the current article as 'project-scene' database.

The Collaborative mechanism is implemented through a series of PHP pages, which contain the above algorithms, communicate with the database and restrict the operations by hiding specific command buttons that correspond to the specific operations. The parametric operations take input from the designer within certain input fields and information regarding an operation is passed to the SQL Layer.

SQL Layer. The SQL layer with the use of series of PHP pages and information from the Collaborative Mechanism layer constructs certain SQL statements, which are triggered by the system, and updates the collaborative database (Figure 5.13.).

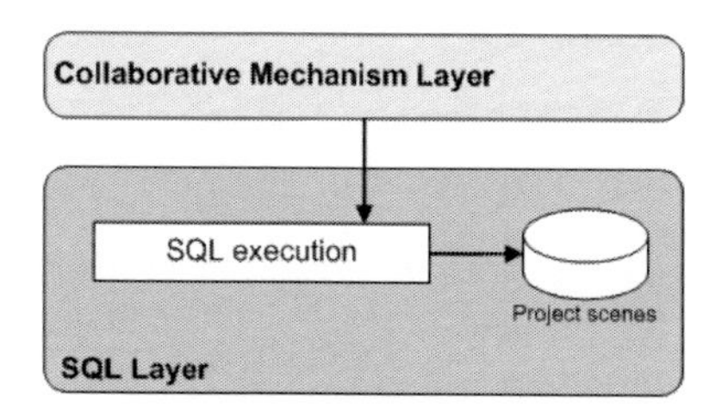

Fig. 5.13. SQL Layer

MultiCAD Client. The client's interface consists of three main sections. The General Information Area, the Main Workspace, and the General Usage command Buttons.

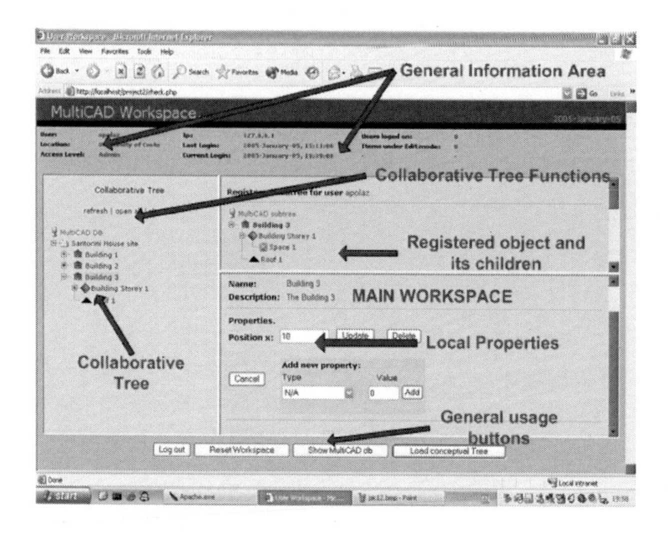

Fig. 5.14. MultiCAD client's browser

The General Information Area contains information regarding the client (designer name, access level, etc). The client can load the PHP script pages in the Main Workspace and use the facilities that they provide in order to use the system. These PHP pages are part of the PHP Script Interface that implements the collaborative environment. The General Usage Buttons are command buttons that control the Main Workspace (Figure 5.14.).

5.4 Case Study

We apply a case study in order to demonstrate the system functionality. We use the system during the early phase of the architectural design process.

5.4.1 Study of Collaborative Activity

Two designers ('idrag' and 'apolaz'), interfere in a collaborative mode for the development of a scene declarative model. Their consequent activities are described as following.

1. The designer 'idrag' locks for editing the object "Building 3" and adds a new object "Building Storey 1" (with a Local Meronymic relation of type: "is_part_of"). Then, he adds one more object "Roof 1" (with a Local Meronymic relation of type: "is_part_of"). Next, the same designer implies between the two objects ('Building Storey_1" and "Roof_1") a new relation (Local Spatial relation of type: Covered_by) (Figure 5.15.).

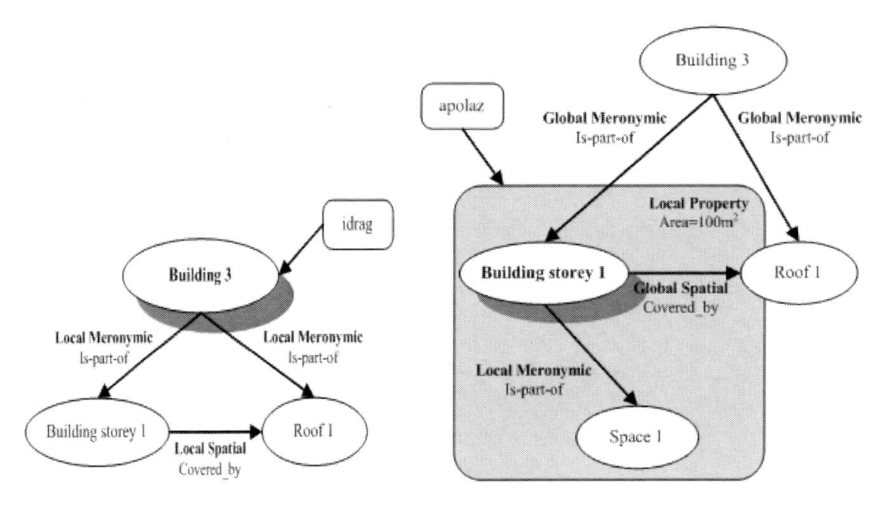

Fig. 5.15. State after idrag's actions

Fig. 5.16. State after apolaz's actions

2. The designer 'idrag' unlocks the object "Building_3", and the designer 'apolaz' locks the object "Building_Storey_1", then he implies the value 100 in the field of the property "Area". Finally he defines a new object "Space_1", within the subtree of "Building_Storey_1", (Figure 5.16.).
3. In the next step, designer 'apolaz' unlocks the object "Building_Storey_1" and locks the object "Space_1". Then, designer 'idrag' locks object "Roof_1". The 'idrag' implies the property "Area" equal to 100 for object "Roof_1". Then designer 'apolaz' imply the property "Length" equal to 200, for object "Space_1", (Figure 5.17.).

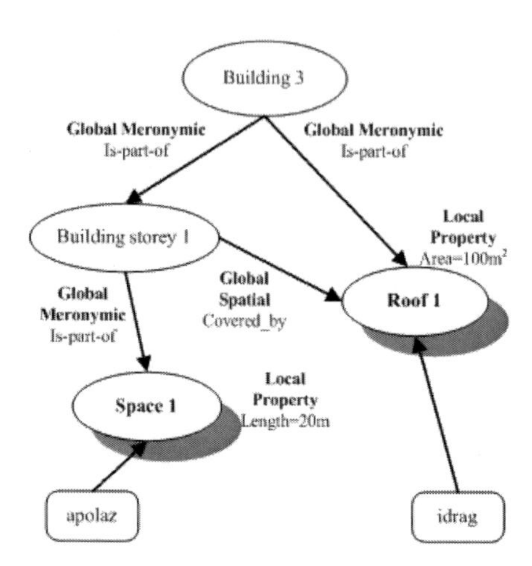

Fig. 5.17. Final state

5.5 Team Profile Module

In the MultiCAD, we have incorporate a team profile module, which is able to estimate a designers team preferences and thus to filter the retrieved solutions according to the designers' information needs. Filtering of the retrieved solutions is a very important task for the MultiCAD architecture and in general for the declarative collaborative environments. This is mainly due to the fact that the scene description is in general abstract and intuitive resulting in multiple interpretations and therefore the retrieved solutions are too many ranging from a few hundreds to hundreds of thousands in typical case. To avoid such unpleasant scenario, designer profile algorithms are incorporated in the MultiCAD architecture. These algorithms are able to filter the solutions according to the current designers' information needs. The module captures the designers' profile through the application of intelligent profile estimation algorithms. Two different scenarios can be discriminated; the single and the collaborative approach. In the single approach, the profile of a designer is estimate independently from the other designers. Instead, in the collaborative algorithm, the profile of a designer is estimated by taking into account the behaviour of all designers participating in the collaborative design. It is anticipated that the second approach provides better performance compared to the first one.

5.5.1 Single Designer Approach

In this section, we describe the algorithm used for estimating the profile of a designer independently from the other. In particular, we assume that the designer interacts with the system and selects a set of relevant / irrelevant solutions from the retrieved ones. Then, an on-line learning strategy is activated which dynamically updates the collaborative system response in order to fit the designer's preferences. In this way, the generated solutions by the system better fit the current designer's information needs. In the proposed MutiCAD architecture, we adopt a intelligent learning schema which is based on a neural network framework. Two different versions of the machine learning algorithm have been implemented. The first uses a conventional feedforward neural network structure for the designer profile estimation. On the contrary, the second approach is based on an adaptable neural network mechanism. The drawback of the first schema is that it requires a large computational effort for designer profile updating. Instead, the second approach is very computationally efficient. Since the focus is set on the collaborative aspect of declarative modelling, we operate here under a certain set of assumptions. In particular, the fundamental assumptions of our approach for the single designer profile estimation are the following:

- All classes are adequately represented in each training set.
- Adapting the neural network to the knowledge implied by a new training set requires small perturbation of the neural network weights.

These assumptions allow application of the mechanisms presented in the following. A complete coverage of the issues that arise with respect to intelligent personalization in a declarative modelling environment, in a wider scope than that of the aforementioned assumptions, is given in the current volume in the corresponding Intelligent Personalization chapter. The latter addresses all aspects of concern for single designer profile estimation, such as the degree of designer participation, the impact of imbalanced datasets, etc.

5.5.1.1 Intelligent Profile Estimation

Every time the designer interacts with the system, he/she selects a set of relevant irrelevant solutions that satisfy his/her information needs. These solutions form a training set S_c which is used for the training a neural network model. In our approach, the neural network approximates the degree of relevance of an object in the database with respect to the scene description. Therefore, modifying the similarity metric, we can achieve a personalized retrieval mechanism where only the solutions that fit to the designer's preferences are first returned. Let us denote as $f(\cdot)$ the non-linear similarity metric which models the relevance degree of an object with the scene description. We also assume that every object in the database are described using a set of properties, as we have stated in the previous sections. Let us denote as $\mathbf{p}_i$ the feature vector that contains of the properties of the object i. In a similar way, we denote as $\mathbf{p}_q$ the characteristics of the scene as they have been provided through the scene description phase. In this way, the non-linear metric is given by the following equation

$$r = f(\mathbf{p}_i, \mathbf{p}_q) \tag{1}$$

In previous equation, scalar r represents the relevant degree among the ith object in the database with respect to the current scene description. The main difficulty, in estimating the non-linear similarity metric is that function $f(\cdot)$ is actually unknown. In order to address this drawback, we adopt in this paper a neural network model, which are generic function approximations [16]. Assuming, that the neural network model is characterized by a set of weights $\mathbf{w}$, we have that

$$r \approx f_{\mathbf{w}}(\mathbf{p}_i, \mathbf{p}_q) \tag{2}$$

where in equation (2), variable $f_{\mathbf{w}}()$ denotes the neural network approximation of the non-linear similarity metric with weights of $\mathbf{w}$. Using equation (2), estimation of the similarity metric $f(\cdot)$ is equivalent to the estimation of the unknown weights $\mathbf{w}$. In our approach the backpropagation algorithm is used for the estimation of the weights $\mathbf{w}$ [16]. The backpropagation algorithm is a steepest descent approach for the estimation of the unknown weights $\mathbf{w}$ which minimizes the least square error

$$E = \sum_{j} (f_{\mathbf{w}}(\mathbf{p}_j, \mathbf{p}_q) - d_j)^2, \ j \in S_c \tag{3}$$

In equation (3), scalar d_j expresses the degree of relevance assigned by a single designer to the retrieved solution *j*, while the $f_{\mathbf{w}}(\mathbf{p}_j, \mathbf{p}_q)$ is the neural network approximation. The summation is accomplished for all selected objects in the training set S_c. The advantage of the proposed method is that is very reliable since it models designers' profile in a non-linear way. However, the main drawback is that it is computationally expensive. To avoid these shortcomings an innovative framework is proposed in the following section.

5.5.1.2 Recursive Implementation

In this section, we extend the aforementioned ideas by introducing a dynamic training framework which is able to recursively estimate the non-linear similarity metric $f_{\mathbf{w}}(\mathbf{p}_j, \mathbf{p}_q)$. In order to do this, we initially assume that a small perturbation of the neural network weights is sufficient to adapt the similarity metric to a new state. Small perturbation of the weights does not imply that deviation of the neural network output before an after the adaptation will be small. This is due to the fact that neural network approximates a highly non-linear function. Under this assumption we have that

$$\mathbf{w}(n) = \mathbf{w}(n-1) + d\mathbf{w} \tag{4}$$

where in equation (4), the variable $\mathbf{w}(n)$ indicate the neural network weights at the iteration *n* of the recursive implementation. Variable ***d*W** denotes the small perturbation amount. Using equation (4) and exploiting the structure of a feedforward neural network architecture, we can derive in analytical relations for the estimation of the small perturbation ***d*W**. Then, since the previous weights, $\mathbf{w}(n-1)$ are known, we can estimate the current weight values $\mathbf{w}(n)$ through the application of equation (4).

More specifically, it can be shown that the small perturbation $d\mathbf{w}$ is provided by the following equation,

$$b = \mathbf{a}^T d\mathbf{w} \tag{5}$$

where scalar b expresses the difference between the output of the neural network model, at the previous weight values $\mathbf{w}(n-1)$ and the target output as has been provided by the designer

$$b = f_{\mathbf{w}(n-1)}(p_i, p_q) - d_i \tag{6}$$

In a similar way, the values of vector $\boldsymbol{a}^T$ depend only from the previous weights $\mathbf{w}(n-1)$ and the structure of the neural network architecture. More details about the structure of the vector $\boldsymbol{a}^T$ can be found in [10]. Equation (5) has been expressed for all selected items by the designer. Assuming that the designer selects a set of relevant/irrelevant data, equation (5) can be easily expressed as a set of linear equation system. Therefore, we have that

$$\mathbf{b} = \mathbf{A}d\mathbf{w} \tag{7}$$

In equation (7), $\mathbf{b}$ is a vector which contains all the differences expressed by equation (6) and $\mathbf{A}$ a matrix that includes all the vectors $\mathbf{a}^T$ for all selecting data. In case that the number of selected items is greater or equal that the number of neural network weights, equation (7) is sufficient for an effective estimation of the small perturbation $d\mathbf{w}$. In particular, the value of the small perturbation $d\mathbf{w}$ is given by the following equation

$$d\mathbf{w} = pseudo(\mathbf{A})\mathbf{b} \tag{8}$$

where the operator $pseudo()$ refers to the pseudo inverse of the matrix $\mathbf{A}$. It is clear that $pseudo()$ is given by the following equation

$$pseudo(\mathbf{A}) = (\mathbf{A}^T\mathbf{A})^{-1}\mathbf{A}^T \tag{9}$$

In case that the set of relevant /irrelevant data are smaller that the number of neural network weights an alternative approach is presented. This is due to the fact that in this case, equation (8) provides no reliable results. More specifically, in this case, we estimate among all possible solutions the one that minimizes the square norm of the small perturbation, that is $\|d\mathbf{w}\|_2 = d\mathbf{w}^T \cdot d\mathbf{w}$. In other words the equations in this particular case are written as

$$\text{Min } \|d\mathbf{w}\|_2 = d\mathbf{w}^T \cdot d\mathbf{w} \tag{10}$$

Subject to

$$\mathbf{b} = \mathbf{A}d\mathbf{w} \tag{11}$$

Using equations (10) and (11), we can estimate the small perturbation of the neural network weights in case that the number of equations, expressed through (7), is smaller than the number of neural network weights. More specifically, it can be easily proved that

$$d\mathbf{w} = (\mathbf{A}^T \cdot \mathbf{A})^{-1}\mathbf{A} \cdot \mathbf{b} \tag{12}$$

The main advantage of the proposed recursive algorithm is that it is computationally efficient. In addition, the solution can be found analytically and thus the issues regarding the trap of the minimization process to local minima are eliminated. It is clear that the form of equations (8) and (12) are simple and analytical regarding the estimation of the small weight perturbation $d\mathbf{w}$ and consequently the estimation of the new neural network weights $\mathbf{w}(n)$.

5.5.2 Collaborative Approach

In the previous section, we have described an efficient non-linear method for estimating the preferences of single designers under a set of assumptions. In the previous approach, we have considered that each designer is acting independently from each other. However, in case of a collaborative environment, it is more preferable the estimation of the designers' profile to accomplish in a collaborative framework by taking into consideration the preferences of all participated designers. Two different approximations have been implemented in the MultiCAD architecture regarding collaborative designer profile estimation. The first scheme is based on a preference consensus algorithm, while the second is relied on a designer profile classification schema through the application of a spectral clustering method [29].

Figure 5.18. illustrates the architecture of the designer profile module. More specifically, Figure 5.18.(a) presents the concept of designer interaction with the system, who evaluates the retrieved results in order to update the system response according to his/her information needs. Instead, Figure 5.18.(b) presents the two main leaning strategies, we have implemented in the scope of this chapter. The first is the single designer profile estimation, while the second is the collaborative framework. As is observed initially, we have a single designer profile activation. This is accomplished in our MultiCAD architecture using the algorithms described above. In particular, we have the recursive approach for the training of the neural network which is described in the previous section. Then, the single designer profile estimation is fed to the collaborative designer profiling algorithms. In particular, two different approaches are implemented. The preference consensus approach and the spectral clustering method. The first method is simple but very efficient, while the second approach yields better results in the expense of complication.

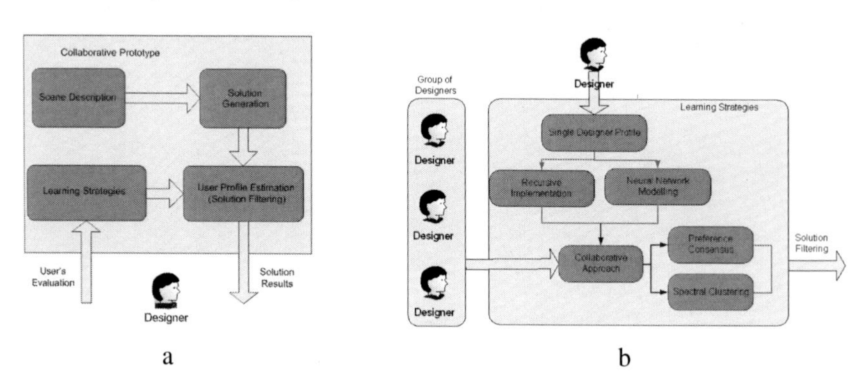

Fig. 5.18. (a) Architecture of designer profile estimation. (b) Implemented learning strategies.

5.5.2.1 Preference Consensus Module

Let us assume that k professionals are involved in the same design process. This means that k different sets of solutions are generated each of which is appropriate for a designers' preferences. It is clear that the k generated solutions are obtained using the algorithm described in the previous section, where each designer is handled independently one from the other. Then, the question that arises is how to combine the k different subsets into a unique set of solutions expressing the best possible consensus for all professionals involved in the specific design process. Let us denote as N_i, i=1,2,…,k a set which contains all solutions that fit the preferences of the i-th designer. Let us, then, construct the union set $N = \bigcup_{i=1}^{k} N_i$ which contains all the solutions that fit the preferences for all designers. As far as the preferences consensus is concerned, two different scenarios can be discriminated; the un-weighted and the weighted scenario. In the un-weighted approach, we assume that all designers, involved in a particular design, have the same importance. On the contrary, in the weighted approach, different priorities are assigned to the designers.

Un-weighted Version: We assume that an occurrence number, say $\boldsymbol{q_i}$ is assigned to each generated solution. This number $\boldsymbol{q_i}$ expresses the times that this particular i-th solution appears in the expanded set N. It is clear that, high values corresponds to "famous solution", that is to solutions which are accepted by most of the designers. On the other hand, low $\boldsymbol{q_i}$ values refer to solutions that are preferable by a small set of designers. Since we have assumed that k designers are involved in the process, the minimum value of the occurrence number $\boldsymbol{q_i}$ is one (only one designer has selected this solution as relevant), while its maximum value is k (all the k involved designers consider this solution as relevant). Since we are in the un-weighted case, in which all designers present the same importance in collaboration, an appropriate policy for selecting the consensus solutions is to select the ones that present occurrence number greater than a specific threshold, say T. In other words, if we denote as C the consensus set, i.e., the set that contains all the feasible common solutions, then, its elements $c_i \in C$, should satisfy the following eqquation

$$c_i \in C: \quad p_i > T \cdot k \tag{13}$$

Threshold T expresses how strict we are in selecting a solution to be included in the final common set. As threshold T increases the strictness in selecting solutions for the final common set increases.

Weighted Version: Let us denote h_i, i=1,.2,..,k the degree of importance that are assigned to a specific i-th designer. This importance corresponds, for example, to the case that some designers may be seniors and thus their contribution should be taken more into account than some other designers. In this case, we modify the occurrence number by multiplying the occurrence of a solution by the importance of the designer, who has selected this solution. Consequently, the minimum number of occurrence number q_i coincides with the minimum value of degree of importance h_i, that is $q_i^{\min} = \min h_i$. On the contrary, the maximum number of the occurrence number q_i will be equal to the sum of all weights, that is $q_i^{\max} = \sum_i h_i$. In a similar way to the un-weighted case, a threshold T is used for constructing the consensus set C. Thus, regarding the policy adopted, for the elements of set C, $c_i \in C$, the following relation is satisfied

$$c_i \in C: \quad q_i > T \cdot \sum_{i=1}^{k} h_i \tag{14}$$

Again, threshold T expresses the strictness in selecting a common solution as in the un-weighted version and can be interpreted as the distance we allow a solution to have from the optimal case, where it is selected by all designers.

5.5.2.2 Collaborative Clustering

The main drawback of the aforementioned scheme is that it does not exploit the dependences among the designers who collaborate together for a design. Instead, the collaborative designers' profile estimation is accomplished based on simple preferences consensus approach. A more sophisticated algorithm is presented in this section, which exploits the dependencies between the designers. These dependencies lead to balanced clusters through the application of spectral clustering methods. In the following, we exploit the constructed clusters in order to get personalized solution in a collaborative environment.

Problem Formulation: Let us consider a vector that expresses the profile of a designer. We denote this vector as $\mathbf{u}_i$ .Vector $\mathbf{u}_i$ includes the degree of relevance of the *i-th* designers with respect to the stored objects in the database. Using the designer profile vectors $\mathbf{u}_i$, we can estimate the degree of similarity between two designers as the correlation degree between the vectors $\mathbf{u}_i$ and $\mathbf{u}_j$. Therefore, we have that

$$p_{ij} = \frac{\mathbf{u}_i^T \cdot \mathbf{u}_j}{\sqrt{\mathbf{u}_i^T \cdot \mathbf{u}_i}\sqrt{\mathbf{u}_j^T \cdot \mathbf{u}_j}} \tag{15}$$

Equation (15) expresses the dependence between two designers, the i-th and the j-th. Using equation (15), we can formulate a matrix which contains all the correlations p_{ij} . We denote this matrix as $\mathbf{Q}$, that is

$$\mathbf{Q} = [p_{ij}] \tag{16}$$

Let us denote as C_r a set which contains designers of the same profile. Then, we can define a metric which expresses the connectives of one cluster,

$$F_r = \frac{\sum\limits_{i \in C_r, j \notin C_r} p_{ij}}{\sum\limits_{i \in C_r, j \in V} p_{ij}} \tag{17}$$

Our purpose is to perform a classification task so that the constructed clusters minimize the quantity Fr. In other world, we have that

$$estimate\ \hat{c}_r : \min F, \ F = \sum_r F_r = \sum_r \frac{\sum\limits_{i \in C_r, j \notin C_r} p_{ij}}{\sum\limits_{i \in C_r, j \in V} p_{ij}} \tag{18}$$

Optimal Clustering: Equation (18) means that we estimate the optimal clusters $\hat{c}_r$ as the ones that provide minimal "flow" to the other clusters. Equation (18) is very difficult to be solved since it is actually a NP-hard problem. To address this difficulty, we write equation (18) in a matrix form. More specification, we denote as **L** a diagonal matrix which contains the sum of the row of matrix **Q**, that is

$$L = diag(\cdots l_i \cdots)\text{, where} \tag{19}$$

$$l_i = \sum_j p_{ij} \tag{20}$$

Then, we can write equation (18) as follows

$$F = \sum_r \frac{e_r^T (\mathbf{L} - \mathbf{Q}) e_r}{e_r^T \mathbf{L} e_r} \tag{21}$$

where vector $\mathbf{e}_r$ is an indicator the elements of which determines the r cluster. In other worlds $\mathbf{e}_r$ has all values of zero apart from the indices that refer to the r cluster, which have value of one. Minimization of equation (21) can be performed using the Ky-Fan theorem [26]. More specifically, the Ky Fan theorem inform us that the optimal cluster, the one that minimize the quantity F , is the generalized eigenvectors of the matrices $(\mathbf{L} - \mathbf{Q})$ and **L** for the fist M eigenvalues. Variable M equals to the predetermined number of the available clusters. However, the generalized eigenvectors of the matrices $(\mathbf{L} - \mathbf{Q})$ and **L** are not the index form. This means that the values are not zeros or ones, as the values of the vector $\mathbf{e}_r$. To round the continuous solution in the form of 0 and 1, we apply to the obtained eigenvectors a k-means classification algorithm. In other words, we classify the generalized eigenvectors to the M available clusters. This classification provides the optimal clusters that simultaneously minimizes the inter-class distances and maximizes the intra-class ones.

Cluster Integration: As representative designer, we select the designer of whom the respective feature vector is located nearest to the centre of the cluster. Let us denote as r_k the representative designer of the cluster k.

After the k representatives have been obtained, each is assigned a weight w_k which is calculated as the sum of two partial weights: a custom index f_k extracted from system use and seniority of the k-th designer, and the density d_k of the k-th cluster which the corresponding designer represents. Formally,

$$w_k = a \cdot f_k + b \cdot d_k \tag{22}$$

where a, b represent constant factors, common for all designers, allowing to fine tune the contribution of each partial weight to the result. The final weights are provided to the system through the *members' importance* vector. In the following we assume that $w_k \geq 1$ without loss of generality.

Finally, for each solution s_i the overall evaluation taking into account the entire group of representatives is calculated as a simple sum, i.e.

$$e_i = \sum_{m=1}^{k} e_{i,m} w_m \tag{23}$$

where $e_{i,m}$ is the evaluation of solution s_i by representative m. Solutions included in the final consensus set fulfil the simple condition of

$$e_i \geq \mathrm{T} \tag{24}$$

where T represents an adjustable threshold for each submitted description, which may be used to control the consensus set population.

5.5.3 Simulations

In this sub-section, we present simulation results as far as the proposed clustering algorithm is concerned and the aforementioned described implementation schemes. Initially, we evaluate the precision measure at different number of retrievals. Figure 5.19. depicts the variation of the precision measure versus the number of the returned solutions. In this figure, we have compared the proposed algorithm with the implementations described in the previous sections for solution filtering.

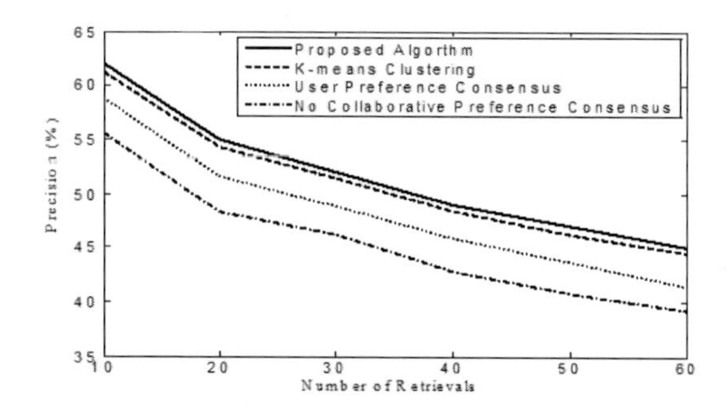

Fig. 5.19. Precision versus the number of solutions returned to the end designer for the proposed algorithm and three other implementations

As is expected, as the number of the returned solutions increases, the precision accuracy decreases. However, the proposed method outperforms the other three approaches providing better precision results at a given number of solution-retrievals. The worst method is the one where no designer collaboration is taken into consideration. Figure 5.20. presents the precision-recall curve for the proposed method and the other three implementations. Again, we observe that our scheme **outperforms** the examined ones. The worst performance is also noticed for the single designer method described in the current chapter, where no designer collaboration is taken into account for the solution filtering.

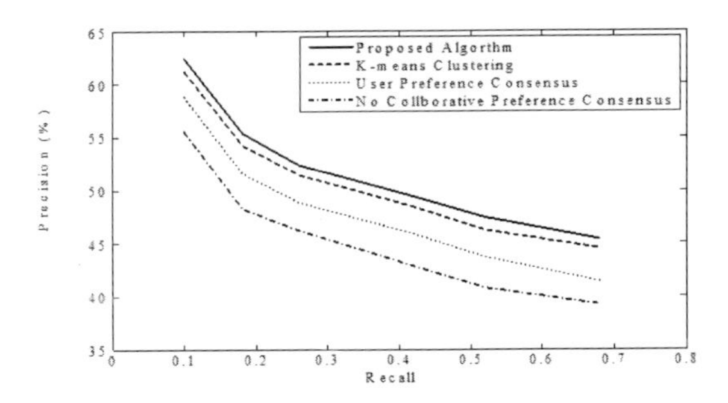

Fig. 5.20. The Precision-Recall curve for the proposed method and other three implementations

Table 5.4. The SER ratio of the proposed scheme compared with other approaches

Average Normalized Modified Retrieval Rank- ANMRR	
The Proposed Algorithm	0.18
K-means Clustering	0.23
Designer Preference Consensus	0.34
Non-Collaborative Preference Consensus	0.45

Table 5.4 presents the ANMRR values for the proposed algorithm and the three different implementation approaches presented in section 9.6.1, i.e., the *k-means clustering*, the *designer preference consensus* and the *non-collaborative preference consensus* method. As we observed, the best performance (smallest ANMRR value) has the proposed algorithm, while the worst has the method where the designer profile is estimated with non collaborative way. An efficient method for *single* designer profile estimation, operating under a less restrictive set of assumptions and a wider scope of system functionality is presented in [4]. Finally, Table 5.4 presents the precision values for different number of clusters. As we observed, small number of designer profile partitions deteriorates the results. This is due to the fact that in this case, we group together designers of different entities.

5.6 Conclusions

In this chapter a complete Collaborative Declarative Module is designed based on CDMS-MultiCAD system architecture and a prototype system is presented in order to evaluate the feasibility and functionality of the CDM framework. This prototype system has been developed based on: PHP language for script interface, MySQL RDBMS for management and storage of MultiCAD databases and JavaScript tools incorporated in interactive pages for the client workspace.

This prototype system was evaluated using four typical cases of habitation [21] which reveal the functionalities and maintain operations of the proposed architecture. In particular, the specific collaborative design environment gives to a designer functionalities of i) working in a personal schedule ii) handling projects even if they are remotely distributed on heterogeneous platforms, iii) reviewing the design work of his/her colleague(s), iv) adopting new design practices, and v) confronting the emergence of a problem within a team.

In the same chapter, we have also proposed efficient mechanisms able to estimate the profile of a designer in order to filter the retrieved solutions according to the current designer's information needs. Two different approaches are discussed; the single and the collaborative approach. The single approach estimates designer's preferences without taking into account the dependencies between two or more designers. On the contrary, the second approach takes into account the designers dependencies resulting in a collaborative framework for estimating the designers' profile.

More specifically, the first (single) approach is implemented through a neural network architecture. Two different learning strategies are investigated. The first trains the neural network architecture without taking into consideration the previous network knowledge. Instead, the second scheme implements a recursive neural network training strategy. Neural network models advances compared to linear classifiers due to their ability to approximate the highly non-linear relationships involved in the preferences of a designer. However, the main drawback of conventional neural network architectures is that their training is not computationally efficient. In addition, the training process may be trapped into local minima, deteriorating network performance. Instead, the second training approach, which is based on a recursive neural network learning strategy, faces the aforementioned difficulties, resulting in a computationally efficient and robust training strategy.

The single (non-collaborative) approach is then exploited towards a collaborative framework. In this case, we take into consideration the dependencies among thc different designers involved in a particular design. The collaborative approach yields a more precise estimation of the designers' profile than the single frame-work. In the collaborative approach, two different scenarios are investigated. The first is based on a simple preference-consensus algorithm, while the second on the use of well separable clustering techniques like the spectral clustering algorithm. Spectral clustering is an algorithm which maximizes the intra cluster correlation, while simultaneous the inter cluster correlation is minimized.

The evaluation has proved the efficiency and effectiveness of the proposed prototype system. Future work concerns further development of the client workspace in order to incorporate different additional collaborative models such as formal language-based descriptions. Furthermore, we work on a full integration of the Collaboration Declarative Modelling System, within MultiCAD type declarative modellers.

References

[1] Anupam, V., Bajaj, C.: SHASTRA – An Architecture for Development of Collaboration Applications. Intern. Journal of Intelligent and Cooperative Information Systems, 155–172 (1994)

[2] Bajaj, C., Cutchin, S., Morgia, C., Paoluzzi, A., Pascucci, V.: Web Based Collaborative CAAD (poster). In: 5th ACM Symposium on Solid Modeling and Applications, Ann Arbor, Michigan, USA, pp. 326–327 (1999)

[3] Bardis, G.: Machine Learning and Decision Support for Declarative Scene Modelling / Ap-prentissage et aide à la décision pour la modélisation déclarative de scènes (bilingual), Thèse de Doctorat, Université de Limoges, France (2004) (2006)

[4] Bardis, G., Miaoulis, G., Plemenos, D.: Handling Imbalanced Datasets of Designer Preferences in a Declarative Scene Modelling Environment. In: Proceedings of the International Conference on Computer Graphics and Artificial Intelligence (3IA), pp. 87–96 (2008)

[5] Bidarra, R., Van Den Berg, E., Bronsvoort, W.F.: Web-based Collaborative Feature Modeling. In: Proc. of 6th Symposium on Solid Modeling and Applications, pp. 319–320 (2001)

[6] Chan, S., Wong, M.: Collaborative Solid Modelling on the WWW. In: Proceedings of the ACM Symposium on Applied Computing, San Antonio, Texas, USA, pp. 598–602 (1999)

[7] Choo, W.C., Detlor, B., Turnbull, D.: Web Work: Information Seeking and Knowledge Work on the World Wide Web. Kluwer Academic Publishers, The Netherlands (2000)

[8] Cross, N., Cross, A.C.: Observations of teamwork and social processes in design. Design Studies 16, 143–170 (1995)

[9] Dietrich, U., von Lukas, U., Morche, I.: Cooperative modeling with TOBACO. In: Proceedings of the TeamCAD 1997 Workshop on Collaborative Design, Atlanta, USA, pp. 115–122 (1997)

[10] Doulamis, A.D., Doulamis, N.D., Kollias, S.D.: On Line Retrainable Neural Networks: Improving the Performance of Neural Network in Image Analysis problems. IEEE Trans. on Neural Networks 11(1), 137–155 (2000)

[11] Dragonas, J.: Collaborative Declarative Modelling / Modelisation Declarative Collaborative (bilingual), Thèse de Doctorat, Université de Limoges, France (2006)

[12] Fribault, P.: Modelisation Declarative d' Espaces Habitables, PhD dissertation, University of Limoges France (2003)

[13] Gero, J.S., McNeill, T.: An approach to the analysis of design protocols. Design Studies 19(1), 21–61 (1998)

[14] Golfinopoulos, V., Dragonas, J., Miaoulis, G., Plemenos, D.: Declarative design in collaborative environment. In: International Conference 3IA, Limoges, France (2004)

[15] Golfinopoulos, V.: Etude et Réalisation d'un Système de Retro-Conception base sur la Connaissance pour la Modélisation Déclarative de Scènes. PhD thesis (In english and french), Université de Limoges, France (2006)

[16] Haykin, S.: Neural Networks: A Comprehensive Foundation. Macmillan, New York (1994)

[17] Kvan, T.: Collaborative design: What is it? Autom. in Construction 9(4), 409–415 (2000)

[18] Lee, J.Y., Han, S.B., Kim, H., Park, S.B.: Network-Centric Feature-Based Modeling. In: 7th Pacific Conference on Computer Graphics and Applications, Seoul, Corea (1999)

[19] Maher, M.L., Cicognani, A., Simoff, S.J.: An experimental study of computer mediated collaborative design. Int. J. Des. Comput. 1(6) (1998)

[20] Makris, D., Ravani, I., Miaoulis, G., Skourlas, C., Fribault, P., Plemenos, D.: Towards a domain-specific knowledge intelligent information system for Computer-Aided Architectural Design. In: International Conference 3IA 2003, Limoges, France (2003)

[21] Makris, D.: Study and Realisation of a Declarative System for Modelling and Generation of Style with Genetic Algorithms. Application in Architectural Design (In english and french). Thèse de Doctorat, Université de Limoges, France (2005)

[22] Miaoulis, G.: Contribution à l'étude des Systèmes d'Information Multimédia et Intelligent dédiés à la Conception Déclarative Assistée par l'Ordinateur Le projet MultiCAD. PhD Dissertation, University of Limoges, France (2002)

[23] Miaoulis, G., Plemenos, D.: Propositions pour un système d'information multimédia intelligent dédié à la CAO – Le projet MultiCAD. Rapport de recherche MSI 96-03, Université de Limoges, France (1996)
[24] Miaoulis, G., Plemenos, D., Skourlas, C.: MultiCAD Database: Toward a unified data and knowledge representation for database scene modelling. In: International Conference 3IA, Limoges, France (2000)
[25] Molina, A., Al-Ashaab, A.H., Ellis, T.I.A., Young, R.I.M., Bell, R.: A review of computer-aided simultaneous engineering systems. Research in Engineering Design 7(1), 38–63 (1995)
[26] Nakic, I., Veselic, K.: Wielandt and Ky-Fan Theorem for Matrix Pairs. Linear Algebra and its Applications 369(17), 73–77 (2003)
[27] Nam, T.-J., Wright, D.K.: CollIDE: A Shared 3D Workspace for CAD. In: Proceedings of the 4th EATA International Conference on Networking Entities, Leeds, UK (1998)
[28] Nam, T.-J., Wright, D.K.: The Development and Evaluation of Syco3D: A Real-Time Collaborative 3D CAD System. Design Studies 22(6), 557–582 (2001)
[29] Ng, A.Y., Jordan, M.I., Weiss, Y.: On spectral clustering: analysis and an algorithm. Advances in Neural Information Processing Systems 14, 849–856 (2002)
[30] Plemenos, D.: Contribution à l'étude et au développement des techniques de modélisation, génération et visualisation de scènes – Le projet MultiFormes. PhD dissertation (1991)
[31] Plemenos, D.: Declarative modelling by hierarchical decomposition. The actual state of the MultiFormes project. In: International Conference GraphiCon 1995, St Petersburg Russia (1995)
[32] Plemenos, D., Miaoulis, G., Vassilas, N.: Machine Learning for a General Purpose Declarative Scene Modeller. In: International Conference on Computer Graphics and Vision (GraphiCon), Nizhny Novgorod, Russia (2002)
[33] Plemenos, D., Tamine, K.: Increasing the efficiency of declarative modelling. Constraint evaluation for the hierarchical decomposition approach. In: International Conference WSCG 1997, Plzen, Czech Republic (1997)
[34] Ravani, I., Makris, D., Miaoulis, G., Plemenos, D.: Concept-Based Declarative Description Subsystem for Computer-Aided Declarative Design. In: International Conference 3IA 2004, Limoges, France (2004)
[35] Rodriguez, K., Al-Ashaab, A.: A review of internet based collaborative product development systems. In: International Conference on Concurrent Engineering: Research and Applications, Cranfield, UK (2002)
[36] Roseman, M., Greenberg, S.: GroupKit: A Groupware Toolkit for Building Real-Time Conferencing Applications. In: Proceedings of ACM Conference on Computer Supported Cooperative Work, pp. 43–50 (1992)
[37] Shen, W.: Web-based infrastructure for collaborative product design: an overview. In: 6th International Conference on Computer Supported Cooperative Work in Design, Hong Kong, China, pp. 239–244 (2000)
[38] Stork, A., Jasnoch, U.: A Collaborative Engineering Environment. In: Proceedings of the TeamCAD 1997 Workshop on Collaborative Design, Atlanta, USA, pp. 25–33 (1997)
[39] Stork, A., von Lukas, U., Schultz, R.: Enhancing a Commercial 3D CAD System by CSCW Functionality for Enabling Cooperative Modelling via WAN. In: Proceedings of the ASME Design Engineering Technical Conferences, Atlanta, USA (1998)
[40] Tay, F.E.H., Roy, A.: CyberCAD: a collaborative approach in 3D-CAD technology in a multimedia-supported environment. Computers in Industry 52(2), 127–145 (2003)

[41] van den Berg, E.: Collaborative Modelling Systems. TU Delft (1999)
[42] Vassilas, N., Miaoulis, G., Chronopoulos, D., Constandinidis, E., Ravani, I., Makris, D., Plemenos, D.: MultiCAD-GA: A System for the Design of 3D Forms Based on Genetic Algo-rithms and Human Evaluation. In: Vlahavas, I., Spyropoulos, C. (eds.) Methods and Applications of Artificial Intelligence, Second Hellenic Conference on AI, Thessaloniki, Greece. LNCS (LNAI), pp. 203–214. Springer, Heidelberg (2002)
[43] Wang, L., Shen, W., Xie, H., Neelamkavil, J., Pardasani, A.: Collaborative conceptual design – state of the art and future trends. Computer – Aided Design 34, 981–996 (2002)

6

Aesthetic – Aided Intelligent 3D Scene Synthesis

Dimitrios Makris

Department of Computer Science, Technological Education Institute of Athens, Ag.Spyridonos St., 122 10 Egaleo, Greece
demak@teiath.gr

Abstract. The synthesis of 3D scenes is a very complicated task in computer graphics. In general, there is a lack of systems that would enable users to utilise aesthetic aspects during the early stages of scene synthesis. The synthesis of aesthetic character of a scene is a complex and intrinsic problem in many design domains. In this chapter, we report a research project that supports the hypothesis that it is possible to encode aesthetic intent within a Declarative design environment in order to aid 3D scene synthesis. The aesthetic intent will have the form of stylistic principles. The aim of this project is to provide designers with a set of solutions, which are most adaptive to their aesthetic intent, in our case stylistic criteria. In order to achieve this aim we define the following objectives. First, the desired stylistic criteria are encoded. Second, they are applied to the Declarative design cycle with the aid of an evolutionary algorithm. The resulting system focuses to aid designers during the early phase of design of 3D scenes. In order to accomplish the aforementioned objectives, within the declarative modelling paradigm we study a methodology in the frame of aesthetic measures and evolutionary algorithms. Such methodology implies the development of a prototype evolutionary declarative design system for the emergence of scenes. The development is made according the following steps. First, we define a modelling scheme of the stylistic intent. Second, we introduce a new generative technique within the Declarative Modelling approach, an evolutionary algorithm for the solution generation. In this way, the designer can introduce a personal style and/or use a style in order to generate solutions adapted to that style. As a particular area of interest, we consider the 3D synthesis of buildings. A demonstration of a series of experiments will provide evidence that the resulted method could successfully quantifying aesthetic intent. Such a system is feasible and efficient for the early stage of 3D scene synthesis.

Keywords: Scene Synthesis, Computational Aesthetics, Style, Declarative Modelling, Evolutionary Design.

6.1 Introduction

The synthesis of 3D scenes is a very complicated design task. The synthesis of aesthetic character of a scene is a complex and intrinsic problem in many design domains. In many design practices, the designer's goal is to define a scene that satisfies functional principles and evokes a certain aesthetic character. In this way, the designer's conceptual notion of a scene involves a diverse set of design intent such as ideas, functional constraints, structural condition and aesthetic requirements. Specifically, the synthesis of a 3D scene involves the development of the physical description of an artefact

G. Miaoulis and D. Plemenos (Eds.): Intel. Scene Mod. Information Systems, SCI 181, pp. 153–183.
springerlink.com

subject to an imposed set of given constraints and specifications. A candidate solution can be considered satisfactory if it is well adapted to the designer's notion for it. Many digital systems offer tools for scene definition and manipulation in a restricted way in which a scene can be modelled. The Declarative Modelling (DM) paradigm has the potential to confront many design tasks and it is capable to attenuate drawbacks and limitations of classical geometric modelling [51]. The introduction of the DM paradigm in scene synthesis has significantly aided many design practices. However, a number of critical issues have still to be faced and overcome to move towards an enhanced declarative design process in which the aesthetic intent is defined and guides the solution generation. In particular, the DM paradigm enables the designer to describe a scene in an intuitive manner. In fact, even it is closer to human intuition and may yield acceptable scene representations not originally conceived by the designer, a moderate declarative description might lead to a large number of alternative scene representations. While they are valid in the sense of compliance with the submitted description, yet they are not equally expressing the designer's aesthetic intent. The latter define a set of stylistic criteria that may have a twofold role in the scene synthesis process. First, it confronts complexity in synthesis [56]. Second, it enables the reduction of alternatives in the problem domain [2]. In other words, stylistic criteria have the potential to boundary the decision within a smaller part of the problem space implying a reduction of the solution space [13].

However, only a fraction of the design intent is regularly incorporated in the declarative description. In particular, the designer's conceptual intent is actually encoded by the declarative description through a set of objects – each featuring certain properties – which are interconnected through a number of relations.

Nevertheless, this description may not capture an aesthetic part of the designer's intents. Hence, the solutions produced for the specific scene may or may not satisfy these non-captured intents. Thus, it is necessary to attempt to encode these missing but desired aesthetic characteristics, which are not covered by the declarative description, and apply them to the Solution Generation and Scene Understanding phases, in order to provide the designer with a subset of solutions, which is most representative of them.

6.1.1 Research Scope

In this chapter, we report a research project that supports the hypothesis that it is possible to encode aesthetic intent within a declarative design environment in order to aid 3D scene synthesis. The aesthetic intent will have the form of stylistic principles. The aim of this project is to provide designers with a set of solutions, which are most adaptive to their aesthetic intent, in our case stylistic criteria. In order to achieve this aim we define the following objectives. First, the desired stylistic criteria are encoded. Second, they are applied to the declarative conceptual cycle with the aid of an evolutionary algorithm. The resulting system focuses to aid designers during the early phase of design of 3D scenes.

6.1.2 Proposed Methodology – Contributing Areas

In order to accomplish the aforementioned objectives, within the declarative modelling paradigm we study a methodology in the frame of aesthetic measures and evolutionary

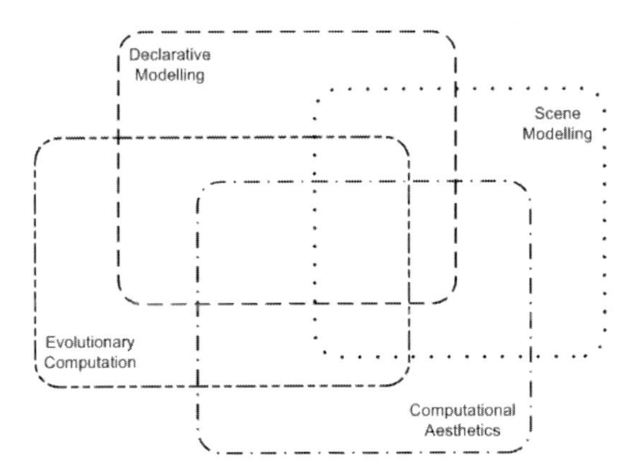

Fig. 6.1. Contributing areas

algorithms. Such methodology implies the development of a prototype evolutionary declarative design system for the emergence of scenes. The development is made according the following steps. First, we define a modelling scheme of the stylistic intent. Second, we introduce a new generative technique within the DM approach, an evolutionary algorithm for the solution generation. In this way, the designer can introduce a personal style and/or use a style in order to generate solutions adapted to that style. As a particular area of interest, we consider the 3D synthesis of buildings. A demonstration of a series of experiments will provide evidence that the resulted method could successfully quantifying aesthetic intent. Such a system is feasible and efficient for the early stage of 3D scene synthesis. The project and the corresponding prototype combine notions from a wide range of fields. These fields are declarative modelling, evolutionary computation, aesthetics / style, and scene modelling. A representation of the interleaving of these fields appears in Figure 6.1.

This chapter is organised as following. In the next section is presented the related work within the areas of aesthetic measures and evolutionary computation. In section 3, we define a general framework of the research approach considering style modelling and multi-objective genetic algorithm within the declarative modelling framework. Section 4 presents the prototype implementation framework. In section 5, we present a design case study that is used for evaluation of the proposed prototype system, with the case of two styles, and we discuss on the efficiency of the applied methodology. Conclusions are drawn in Section 6 concerning the applicability of the model, and its contribution.

6.2 Related Work

In this section, we discuss some relevant effort and approaches on the two overlapped contributing areas of this research. In particular, we will concentrate on evolutionary computation and computational aesthetics. The understanding of design with the aid of aesthetic intents will be a significant advantage in developing advanced Computer Aided Conceptual Design – Computer Aided Design (CACD – CAD) environments.

6.2.1 Evolutionary Computing Techniques

Evolutionary algorithms (EAs) is a category of search algorithms. In general, such algorithms transform a problem following particular conditions of search. In this way, all potential solutions to the problem fill a search-space, where a solution is defined as a point of this space [38]. In the field of design, the task of improving a design is defined as a problem of searching for better solutions across the space of valid designs, a sub-space of search space. Natural evolution paradigm is the origin and the fundamental insight of the evolutionary search algorithms. Such algorithms they perform search by evolving solutions to problems, in a similar way as natural evolution mechanism. In this way, they operate with a large collection of solutions at once, instead of operate with a single solution at a time in the search space. EAs have been performed with high flexibility, efficiency and robustness comparing with all known search algorithms in computer science. As an apparent result they are widely utilised to confront a great variety of diverse problems [55], [46]. In EAs a random or semi-random population of individuals, (a candidate solution of the problem), evolves iteratively while a set of stochastic operators known as selection, recombination and mutation is periodically applied on it. The population of individuals strive for survival: a selection scheme, biased towards fitter individuals, selects the next generation. After a number of generations, the algorithm converges guided by selection based on the predefined fitness function or diverges with random selection. This process results to find global optimal and/or adapted solutions to the design problem. The best-known EAs included: Genetic Algorithms, Genetic Programming, Evolutionary Programming and Evolution Strategies.

Genetic Algorithms (GAs) is a search technique adequate for searching noisy solution spaces with local and global minima [35]. The probability of the search getting trapped in a local minimum is limited, because it searches from a population of points, not a single point. GAs start searching by randomly sampling within the solution space, and then use stochastic operators to direct a 'hill-climbing' process based on objective function values [30]. Despite their apparent simplicity, GAs have proved to have high efficacy in solving complex problems which other, more conventional optimisation methods, may have difficulties with, namely by being trapped in local minima.

Genetic programming (GP) is an evolutionary approach for programs generation [40]. GP induces a population of computer programs that is improved automatically as they experience the data on which they are trained. GP techniques, as conventional GAs, involve initialisation of a random population of computer programs, and for a specific number of generations, evaluating the fitness of each individual program according to some fitness functions, and then applying some genetic operators. One of the advantages of GP is that it uses a variable length genome, which reflects hierarchical and dynamical aspects of the potential 'solution(s)' to a particular 'problem'. The growth of GP systems provides great potential in geometric modelling and design problems, [16], [17], [67], [68].

Evolution Strategies (ES) are of two types ES (m, b) and ES $(m + b)$ [58]. In both types m parents produce b offsprings with recombination and mutation operators. In both types of ES each individual has a mutation operator with adaptive mutation rate. However are different in the following. In the ES (m, b), the best b offspring survives and it replace the parents. In this way parents are not included in the next generation.

In the ES $(m + b)$ both parent and offsprings are allowed to survive. The $(m + b)$ ES is based on an elitist strategy while the (m, b) ES is not.

Evolutionary Programming (EP) is similar with ES. An offspring is generated from a parent by uniform probability and adaptive mutation and no recombination at all. Then from the union of parent and offsprings a variant of stochastic tournament selection selects m best individuals. In this way the best individual is always retained providing that once an optimum is found it cannot be vanished [22].

6.2.1.1 Evolutionary Design

The foundation of evolutionary design is the synthesis of the diverse evolutionary computation methods with the design process in a variety of design domains. In general evolutionary computation methods have the ability to perform as general purposed problem solver. Such ability is in analogy with the designer intelligence although it provides faster and efficient results. Many traditional Artificial Intelligence methods have a disadvantage in modelling design intelligence. In many cases it is modelled explicitly as knowledge in representation and inference. The weakness of traditional methods is a result of the little understanding about the way of utilisation of knowledge by designers.

In the design domain, the use of evolutionary search for optimisation and adaptation is becoming widespread. Design problems are ill-defined, nonlinear, dimensional and complex in this way application of evolutionary computing algorithms provide robustness and high adaptive capabilities. Their main advantages of evolutionary computing are the following. First, they do not require explicit knowledge of the problem structure or differentiability. Second, they provide multiple near-optimal solutions to even ill-defined problems. According with the content and the characteristics, evolutionary design is classified in four approaches [6]:

- Evolutionary design optimisation. Optimisation of existing designs by evolving the values of suitably constrained design parameters,
- Creative evolutionary design. Generation of entirely new designs from little abstract knowledge to satisfy functional requirements,
- Conceptual evolutionary design. Production of high level conceptual frameworks of preliminary designs,
- Generative evolutionary design. Production of forms of designs that contributing to the emergence of implicit design concepts.

The evolutionary process of the abovementioned approaches merge many fundamental characteristics of design intelligence, such as design data and information modelling, intuition emergence, concept development, model optimisation and adaptation, qualitative and quantitative evaluation. Very often evolutionary approaches are incorporated in Computer Aided Design/Computer Aided Engineering systems resulting towards intelligent design and decision support systems.

6.2.1.2 Genetic Algorithm Applications in Design

Across diverse areas, different problems have had solutions successfully optimised by GAs. Design cases are one of common problem areas to use GAs as a way of optimisation. GAs are being applied to many areas of engineering design in mechanical engineering, electrical engineering, aerospace engineering, architecture and civil engineering, et cetera It is practically impossible to give a comprehensive overview of all existing applications even for one such area. We have decided to discuss conceptual design, and shape optimisation. The common feature of these areas is their strong geometric nature, which is also important in most design problems. This also indicates that GAs can be efficient in solving problems with very different engineering content within a similar framework and by using similar procedures.

The early phases of the design process contain the conceptual design of an artefact. The designers operate creatively, in general, in two ways, they utilise new concepts, or they blend known concepts in a original manner. During that phase of design process, they also decide about which design parameters are to be optimised. However, the literature provides examples of numerous ways for the creation excellent conceptual designs. Additionally, studies have shown that the computing paradigm of adaptive search techniques with emergent solution characteristics is well suited to the complicated and unstructured nature of the conceptual design process [33]. The following literature review has a particular view on research concerned with conceptual design of buildings, bridge design, spatial planning, product design, and airframe design. However, it is important to note all areas of research are under continual development. Furthermore, it is necessary to mention that the researches discussed in the following pages do not cover all aspects involved in the global conceptual design process. Nevertheless, we have tried to address the problem from several perspectives.

Frazer developed evolutionary design environment with the use of GAs and evolves unpredicted building forms and their possible interactions with the environments [23], [24]. In [7] a design system uses a GA where conceptual designs of 3d solid objects were created and optimised. A multi-objective optimisation allows users to define design problems without fine tuning large numbers of weights. A variable-length chromosome in GAs is addressed to allow variable number of primitive shapes. A new hierarchical crossover operator uses a semantic hierarchy to reference chromosomes. The system evolves conventional and unconventional designs. In [36] employed a GA for space layout planning. In the study they optimise the distribution of available space among different activities in a building in order to minimise the cost of taxiing between those activities. They concluded that a GA is able to generate good designs for complex design problems. In [50] developed a multi-criteria GA for optimal conceptual design of medium-rise buildings. Two conflicting criteria are simultaneously optimised. Specifically, Pareto optimal equal-rank designs non-dominated in both criteria by any other feasible design are found. There is a performance trade-off between the objective criteria and it is up to the designer to make some compromises to arrive at an acceptable design. In [27] it is used GAs to enlarge the state space, so that the set of possible designs changes. They generalise crossover in such a way that it can move the population outside the original state space. This strategy supports creative design, and therefore can be used in the conceptual stage of the design. In [61] discussed a methodology for the evolution of optimum structural shapes. A GA is used to evolve optimum shape designs that are free to assume any geometry and topology and do not resemble any conventional design. The methodology addresses configurationally and topological aspects of the design.

In [47] applies GAs to the design of 3D shapes, during early stages of design mainly by proposing innovative shapes. The difficulty of finding appropriate objective functions for design evaluation have made some of these attempts remains at an experimental level, with selection and crossover being performed manually by the designer. In [19] is explored the problem of conceptual engineering design and the possible use of adaptive search techniques and machine based methods. A GA is adapted to multi–objective optimisation within conceptual design problem. Interactive dynamical constraints in the form of design scenarios are introduced with machine–based agents in conceptual design process. They are integrated with the conceptual engineering design system. In [11] has suggested a GA system for building's environmental performance in relation with the study of building envelope design. The system generates building forms, departing from abstract relationships between design elements and uses adaptation to evolve architectural form. The system provides many diverse options, which might not be 'optimal' but have a good environmental performance and be architecturally interesting at the same time. In [66] in a following project an evolutionary design system is developed in which multi-objective evaluation and selection support the process of automatically generating geometric forms of products using primitive shells that are easy to manufacture. In [69], constraint-programming techniques are used in order to find a set of solutions (3D forms) that satisfy the spatial constraints imposed by the user and create an initial generation. In the sequel, GAs operators generate new solutions and interacts with the user for solution evaluation and increase the speed of convergence to those forms that satisfy his/her aesthetics.

6.2.2 Computational Aesthetic Approaches

The introduction of computing theories in the field of aesthetics, art formation, and design, has its origins during the 1960s. The interweave between these fields was amplified by the application of hardware, software, and cybernetics to provide support in the evaluation and interpretation of art, and in the creation of artefacts. Several researchers have attempted to define the discipline of Computational Aesthetics (CAs) in the field of computer science. The field of CAs is characterised by research and development of computational methods across a variety of application domains such as 3D modelling, computer graphics, product design, interface design, building design, computer aided design and computer aided styling. CAs provide methods and techniques for the quantification of qualified criteria in design and decision-making. There exist many research efforts considering the development of aesthetic measure(s) integrating mathematics, information theory and computers within artistic creation, aesthetic evaluation and interpretation.

The first attempt towards a formalisation of aesthetic measure is the work of Birkhoff [8]. The 'aesthetical measure' is a comparison measure related only to similar objects, objects of a 'family'. Max Bense moves from metaphysical aesthetics through mathematical aesthetics to informational or aesthetics. His work is inspired by Birkhoff, where Bense retained his formula but transformed it into the informational measure: redundance divided by statistical information. Redundance is obviously the same as order by which the elements are connected. In this way, he provides a differentiation of aesthetics in macro-aesthetics and micro-aesthetics. The former is concerned with the evident realms of perceptions of the aesthetic object, while the latter with the not-evident realms of the aesthetic object. [5]. Aesthetic measures applied in many design fields like 2D layout, web site aesthetics, human-computer interface design, jewellery conceptual design, artistic evaluation-creation, bridge design, and product design.

In [34] is presented an aesthetics measure that is used in automated 2D layout. A set of heuristic measures are combined for the features of alignment, regularity, separation, balance, white-space fraction, white-space free flow, proportion, uniformity and page security. In [41] is presented a set of four studies in order to develop a measurement instrument of perceived web site aesthetics. In [45] it is presented an attempt for evaluating interface design. A set of aesthetic characteristics have been tested and the results suggest that these characteristics are important to prospective viewers. In [71] is described an approach that combines a new mathematical model of aesthetic measure by considering theory of aesthetics, Iterated Function Systems fractal's characteristics and human perception on aesthetics. The resulted quantitative aesthetic model is used as fitness function in an evolutionary algorithm for jewellery conceptual design. Staudek [64] developed a system with algorithmic aesthetics, with the integration of computer into artistic creation and aesthetic evaluation. He used aesthetic functions to evaluate aesthetics in terms of order, complexity, harmony, variety, entropy, and redundancy. In [26] it is considered the aesthetic of bridges by a set of aesthetic measures. These measures are based in a set of factors (structural configuration, functional characteristics balance and slenderness) which guide the synthesis of bridge components. In [29] it is developed aesthetic-driven tools for shape retrieval and modification, which allows industrial stylists – designers to optimize product design phases according to aesthetic and engineering requirements of products.

6.2.3 Style Modelling Approaches

In this sub-section, we discuss some relevant effort and approaches on design by style. The understanding of design style will be a great advantage in developing advanced Computer Aided Conceptual Design/Computer Aided Design systems in design [59]. The derivation of procedural knowledge of style is useful in design. Such knowledge enables designer to uncover the source producing style characteristics of products and to use them for producing designs under that style and/or evolve further a style. Recent studies of the explicit representation of style include the style of Taiwanese traditional house style [15], and the style of renaissance architect Palladio [65]. Both approaches imply the description of style in the form of rules of composition for shape grammars. The computer can manipulate these approaches in order to generate designs with that style. Aesthetic and in particular stylistic concepts are very important in every design product. Many researchers underline the significant missing of systemisation and formalisation of aesthetic related design knowledge [10]. However, the main problem it was the fact that same aesthetic properties can be associated to different shape parameters. As a result, it remains difficult to provide an absolute definition of aesthetic character, which makes very inconvenient the explicit linking with aesthetics knowledge.

6.2.3.1 The Concept of Style

We provide two approaches for the concept of style, the object and the process view. The definition of style through an object view depends on the set of common characteristics of products. The emergence of style is based on a repetitive appearance of a set of 'features' in a number of products, such as buildings, paintings, poems, and music. A list of the characteristics of a work of art includes: first convention of form, secondly material, and thirdly technique, [1]. In a similar approach formal and qualitative characteristics are more decisive for the formation of a style, [60]. For that reason, he proposed the following aspects for the description of the style: form elements or motives, form relationships and qualities. An analogous approach appeared in architectural history, [63].

In the process view, the formation of a style is achieved through some important factors that characterise a creative process. This approach is concentrated in the way of doing things. During a process three aspects play an important role, choices, constraints, and search orders implemented in a process [12]. The birth of a style could emerge around a set of choices between ways of performance or procedures during the creative process, [32], [62]. For Gombrich, these choices are of the kind of perceptual cues, while for Simon these choices are those that emerge during the design process. The decisions made among alternatives characterise a style. The second factor is the constraints that were imposed during the creative process. Constraints could strongly affect the selection among choices. The third factor is the search order. Simon argues that human designers provide specific procedures for the priority of goal satisfaction and constraints application [62]. During the design process, the order of satisfaction of certain aspects of design imposes a sequence in the satisfaction of the design aspects. Such a procedure could result in a distinct form and composition for the design artefact.

In a research program [53], [28] academic and industrial partners approach the styling process in product design. Styling is a creative activity where the designer's goal is to define a product that evokes a certain emotion while satisfying obligatory ergonomics and engineering constraints. Their approach enables the development of tools to preserve the aesthetic design intent during the required model modifications. In [14] designers are provided with a language that can communicate stylistic concepts to the computer for form generation. A mechanism for comprehensive formal style analysis could records essential properties of styles, and it could serves as a framework for style knowledge accumulation. Such infrastructure enables the communication of stylistic concepts to computer for form generation of products. In [70], a research is focused on the impact of form features on style through the topological structure and geometrical variation. In particular, they evaluate the influence of design style by feature geometry. They argue that a combination of topological and geometrical attributes on both levels of global and detail perception could enable a definition of a design style. The user applies a set of procedures on stylistic features of an object. In [20] it is approached the interpretation of architectural style using syntax – semantic model. Thus, it is possible to regard a style as a representation of common particular meanings called complex semantics expressed from a set of designs. Style could be analysed into an ensemble of complex semantics as emerged by the synthesis of some simple semantics after the articulation of a group of forms. The choices of the specific forms, and the creation of some new forms, are directed by the design decisions. The notion of simple semantics could derive from properties of forms (geometric) and relationships (topological, organisational) among them. In [49] a project concerns the study of beauty in morphological design based on the assumption that universal aesthetic principles exist and are quantifiable. The resulted computer-based system for quantification of aesthetics in the presence of proportionality is utilised in the analysis and synthesis of various design artefacts. In [54] designers have a tool for the generation of families of artefacts under a specific style during the early stages of product development. In this way, designers through changing structures through generative rules can maintain stylistic consistency. In a project [48], shape grammars (SG) are used for the quantification of differences between vehicle classes, through the application of class-specific rules and constraining rule applications to within parametric ranges determined for each class. As a result, SG enable style to be maintained across and within a family of vehicles. In [37] style is considered as an ordering principle of artefacts in a design domain. In this way artefacts are structured through the visual similarity of its constituent members. In [39] is presented a method for representing and reasoning about the relationship between physical and abstract characteristics. In this way, style is characterised in terms of experiential features that are computed physical features' relations.

6.2.4 MultiCAD Framework

Declarative scene design consists of three sequential phases (Figure 6.2.). The Description phase where the designer describes intuitively a model. The Generation phase during when several solutions are generated that conform to the input description. During the Understanding phase scene solutions are fed-back to the designer (visualisation process) and in turn, they can be evaluated (evaluation process). The MultiCAD framework is a platform for developing multimedia information management systems supporting all phases of the declarative process.

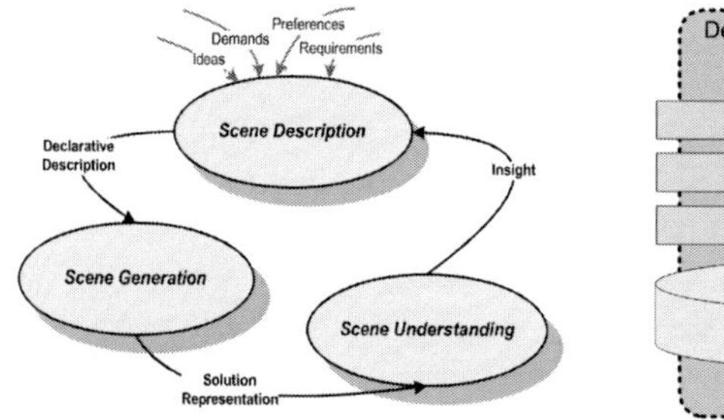

Fig. 6.2. Declarative design cycle

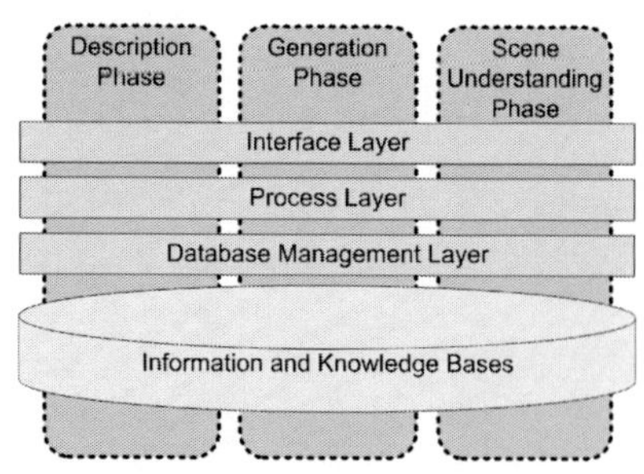

Fig. 6.3. MultiCAD multi-layered architecture

MultiCAD is a multi-layered architecture that comprises three layers (Figure 6.3.), the Interface layer, the Process layer, and the Information and Knowledge management layer. Chapter 1 of the current volume presents their functionality. Figure 6.3. presents their projection to the declarative design cycle phases. There exist different system prototypes based on MultiCAD architecture [4], [21], [25], [31].

6.3 Research Approach

This section is organised in two sub-sections. In the first sub-section we introduce an architectural style scheme within the declarative design system, with the aid of knowledge modelling framework. It is also analysed how it is possible to affect the three phases of the declarative conceptual cycle. The architectural knowledge includes stylistic information for the creation of buildings. Different styles of each design or from different designs could be stored in a Style Library and used for the advantage of future design cases. The adopted notions of architectural style in the proposed structure will further facilitate its incorporation in an evolutionary declarative design system.

In the second section, we present the multi-objective genetic algorithm (MOGA). We focus on the development of a new generation engine that is based on a multi-objective optimisation genetic algorithm within MultiCAD. The evolutionary declarative design environment will direct the development of new designs in two interconnected cycles. The first cycle will concern the evolution of the spatial planning of a building design; the second cycle will concern the evolution of roof morphology of a building design. In parallel, the new generation engine will concern a specific design environment for the user of the system. The user will provide declarative descriptions and with the use of the MOGA will result in design solutions adapted to a particular architectural style. A model of an evolutionary declarative design system is analysed in a specific design context of implementation, that of architectural conceptual design.

6.3.1 Architectural Style Modelling

In this sub-section we develop a model within a declarative design system in order to capture and express qualities of building representations, (relationships - properties), that are important in reasoning with stylistic design principles. In this way the designer will have the ability to decide which features-properties of style to be introduced in a declarative design environment for the adaptation to a particular style, possible evolution of a style and the emergence of new styles.

The main intention is the development of an informationally complete and semantically fertile building's style representation. Such a representation is competent in supporting both the interpretation and the evolution of architectural styles within a declarative design system. We develop a general structure for representing an architectural style that responds to two issues:

- Stylistic information must be represented independently of style category. Style representation must follow a frame, which can be easily adapted among the different architectural styles, and incorporate all different knowledge of styles.
- The representational frame must ensure a full representation of style semantics.

This representation is based on the following two principles. First, we address only one phase in the overall design process. In specific, the focus is on conceptual design phase of architectural design. Second, we integrate the representation of spatial, geometric and structural characteristics in one unified representational frame.

6.3.1.1 Style Knowledge Framework

We employ the distinction in the appearance of constraints in the declarative description of a 3D scene. So far, such a distinction has not appeared in the applications of dedicated declarative modellers. Therefore, first we consider any constraint of the initial scene description as hard constraint that should not be violated during the generation of alternative solutions. Second, the architectural style that the designer prefers for his/her description to adapt as soft constraints. The relations representing soft constraints play an important role for both the representation of style and during the generation of solutions. In the first case, these relations are fundamental for the expression of the principles of an architectural style. In the second case, these soft constraints will play the role of objective criteria for the evaluation of the evolved solutions by the Genetic Algorithm during the generation phase. We will explain how this distinction will be introduced in the Description and in the Generation phase of a declarative modeller such as MultiCAD.

Style representational scheme

We introduce an enriched modelling framework for architectural style that will be based on the architectural design knowledge and the Declarative Knowledge Framework for Architecture-oriented Building Modelling (DKABM) [42] framework. Next, we provide updated and design oriented geometric entities capable to deal with the demanding representation of architectural designs. The layout of an architectural style can be interpreted according the following definition: The layout of a style is a set of its compositional components, represented by abstract geometric elements embodied by the overall shapes and dimensions. The elements of the layout are composed together by means of stylistic principles (in the form of objects and constraints), which determine the architectural character of the layout. The fulfilment of this approach necessitates the introduction of the subsequent description concepts. The abstract elements used for representing spatial and building elements are called geometric entities. There is an obvious need for a representation as a basis that will define comprehensively: first the constituents of topological relationships. Second the constituents of formative conditions, and third the constituents of building typology.

In that scheme, it is of equal importance how the interrelations among the three categories are defined. In the subsequent parts of this section, first we explain the data scheme of the various elements that constitute the above representational scheme, with reference to a simple geometric entity. Second, we apply that scheme to the proposed two example of architectural styles.

Declarative Knowledge Framework for Architecture-oriented Building Modelling

The introduction of style within the MultiCAD system architecture was based on a representation that concerns declarative modelling and architectural knowledge. An approach for uniform scene description is introduced as Declarative Knowledge Framework for Architecture-oriented Building Modelling [42]. The DKABM can cover all required information to describe consistently and completely any building seen from architectural point of view. The description of a building is organised in categories of objects with common properties and relations. Information is organised hierarchically forming a knowledge decomposition tree influenced from the Declarative Modelling by Hierarchical Decomposition [52]. The DKABM framework facilitates the identification of semantic relationships, which is needed to define architecture-oriented building models (Figure 6.4.).

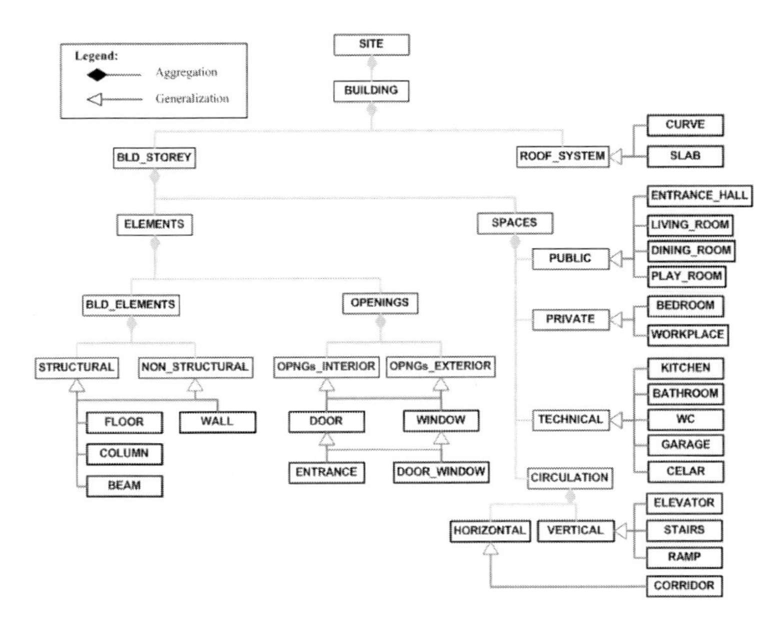

Fig.6.4. The DKBM

A complex scene could be described with the use of constraints. Any relations between objects as well as properties of objects could be specified by the declarative semantic of constraints. However, we will employ an important change on the above approach, because of two reasons. Firstly, during the architectural design process not all constraints are of the same importance. Therefore, the achievement of an architectural design oriented declarative design system it is needed the distinction between constraints. Secondly, in order to represent architectural style it is fundamental to use and manipulate constraints. In particular, we utilise constraints not only for scene description, but also in advance for the expression of architectural style principles. The achievement of such effort makes necessary the distinction of constraints as hard and soft. We adopt the distinction of hard and soft constraints in the Knowledge Database of MultiCAD (Chapter 1). Such separation is provided for the first time in a declarative modeller.

Semantics for architectural style

We present a framework within which semantics is effectively incorporated in the definition of architectural style(s).

- A variety of constraint types is defined, each of which can capture specific aspects-characteristics of stylistic intent (stylistic principles).
- A class of features is defined, integrated with the constraints.
- A formal geometric definition for a semantic stylistic model is presented, providing the basis for a comprehensive definition of different architectural styles. Constraints offer a convenient means to specify properties and characteristics, which the designer requires in a style model.

Stylistic constraints - properties

Three types of constraints are used to specify stylistic characteristics in and among elements of a style: topologic, geometric and algebraic constraints.

- Geometric constraints: geometric relations (e.g. parallelism, perpendicularity, et cetera.) between style elements. Geometric constraints can be used to specify geometric properties-characteristics either within an object instance, or between sets of different object instances. An example of the former is the characteristic 'the side faces of living room and kitchen should be parallel', and an example of the latter is 'the axes of two holes should be at a specified distance'. An essential property of geometric constraints is that they operate on elements of architectural styles, their faces, starting points.
- Algebraic constraints: expressions like equalities or inequalities, among elements parameters. Algebraic constraints can be used to specify parameter relations within an element, e.g. equality between the width and the length of a space, or among parameters of different elements, e.g. 'the length of a living-room should be twice the length of a bedroom'. Additionally algebraic constraints are used for the specification of the allowable range of values for an element parameter, by means of inequality relations.
- Topologic constraints: relations concerning the relative position of stylistic objects (spatial and/ or structural). Topologic constraints are specified on elements of an architectural style by using the topologic relations. Such topologic relations are the following: Near_to, Adjacent_of, Above_of.

Definition of objects

Stylistic objects are defined as representations of form aspects of different parts of a building, and are significant for the definition of an architectural style. Each element has a well-defined meaning, expressed through its topologic, formative and building properties. Form and parameters of elements are specified using stylistic constraints. Each stylistic element could be represented as parameterised object. A basic object encapsulates a set of geometric constraints that relates its parameters to the corresponding object. The geometry of an element accounts for the bounded region of space comprised by its volumetric form. Moreover, its boundary is decomposed into functionally meaningful subsets, its faces. Each face is labelled with its own generic name, to be used in design-compositional operations. For example, a cuboid has a top, bottom, east, west, north and south face.

Spatial planning constraints

Given the abovementioned distinction, we obtain as hard constraints the following binary relations for the arrangement of the spatial objects of a description. Two elements could be adjacent on one of their six faces, (North-South-East-West-Above-Below). In order to enable the circulation between the two spaces we consider a minimum distance of adjacency. There are the following, Adjacent_North, Adjacent_South, Adjacent_East, Adjacent_West, Adjacent_Above, and Adjacent_Below.

Two elements could be near each other on one of their six faces, (North-South-East-West-Above-Below). Furthermore, it is considered a minimum distance of nearness. There are the following, Near_North, Near_South, Near_East and Near_West.

Two elements could overlap on one of their four faces, (North-South-East-West). A minimum variable distance of overlapping could be considered between the two spaces. There are the following, Overlap_North, Overlap_South, Overlap_East, and Overlap_West. A set of six relations that concern object's placement within the site. These are: Place_North-East, Place_North-West, Place_South-East, Place_South-West, and Place_Centre. The relation Inside forces placement of an object within the boundaries of another object.

We obtain as soft constraints a number of binary and unary relations. Considering the geometric properties of the spatial objects we have, the following relations are defined:

- Binary relations: Same_Length, Same_Width, Same_Height, Same_Area, Same_Volume.
- Unary relations: Longer_than_Wide, Longer_than_High, Wider_than_Long, Wider_than_High, Higher_than_Long, Higher_than_Wide.

For the relative placement of the objects we have the following binary relations: Axial-Centre_North-South, Axial-Centre_South-North, Axial-Centre_West-East, Axial-Centre_East-West. Axial-Length_East-South, Axial-Length_East-North, Axial-Length_West-North and Axial-Length_West-South. Axial-Width_North-East, Axial-Width_North-West, Axial-Width_South-East, and Axial-Width_South-West.

The constraint Align is defined as adjusting objects position in relation to a certain line. Parameters for the Align operator are the alignment line and edges of objects to be aligned. Also, objects locations can be adjusted with regard to any part of these elements, i.e. edges, centres, or axis. These are: Align-Length_East-South, Align-Length_East-North, Align-Length_West-North and Align-Length_West-South. Align-Width_North-East. Align-Width_North-West, Align-Width_South-East and Align-Width_South-West. All constraints could be applied both on the building as global constraints, and /or the spaces as local constraints.

We should underline the fact that in many research projects [9], [25] they appeared similar relations. However, their application in our project is differentiated in the following points. Firstly, there exists a separation between hard and soft constraints, rather than utilised as hard constraints in the beginning of the description. Secondly, they are utilised as evaluation criteria. The benefit is that it is possible the expression of a variety of styles with the same soft constraints.

Aesthetic evaluation

In order to capture the aesthetic intentions of the designer we also introduce a number of aesthetic criteria in the form of soft constraints. Such aesthetic criteria also applied for style evaluation: Balance, Equilibrium, Unity, Density, Regularity, Homogeneity, Rhythm, Symmetry, and Proportion [45], [64].

Balance is the distribution of optical weight in a 3D design layout. Optical weight refers to the perception that some objects appear heavier than others do. Larger objects are heavier, whereas small objects are lighter. Balance in 3D layout is achieved by providing an equal weight of objects, left and right, top and bottom. It is computed as the difference between total weighting of objects on each side of the horizontal and vertical axis.

Equilibrium is stabilisation, a midway centre of suspension. It is computed as the difference between the centre of mass of the displayed 3D objects and the physical centre of the bounding box.

Unity is coherence – totality of objects that are visually closely all in one piece. In planning design it is achieved by using similar dimensions and leaving less space between objects of a design than the space left at the margins and within the objects. The objects are dimension-related, grouped together and surrounded by white space, (if no specific initial space layout is defined, then the layout and bounding box is the same).

Density is the extent to which a specific space layout (or the resulting bounding box of a set of objects), is covered with objects.

Regularity is uniformity of 3D objects based on some design principle or plan. In 3D design, it is achieved by establishing standard and consistently spaced horizontal and vertical alignment points for objects, and minimizing the alignment points between them.

Homogeneity. The relative degree of homogeneity of a 3D composition is determined by how evenly the objects are distributed among the four quadrants of the plan. The degree of evenness is a matter of the quadrants that contain more or less nearly equal numbers of objects. It is by definition, a measure of how evenly the objects are distributed among the quadrants.

Rhythm in 3 D design refers to regular patterns of changes in the objects. This order helps to make the appearance exciting. It is accomplished through variation of arrangement, dimension, number and form of the objects. The extent to which rhythm is introduced into a group of elements depends on the complexity.

Symmetry concerns axial duplication: an object on one side of the centre line is exactly replicated on the other side. Vertical symmetry refers to the balanced arrangement of equivalent objects about a vertical axis, and horizontal symmetry about a horizontal axis. By definition, it is the extent to which the spatial layout is symmetrical in two directions: horizontal and vertical.

Proportion. Aesthetically pleasing proportions should be considered for objects In 3d design 'Proportion', is the comparative relationship between the dimensions of the design objects and proportional shapes.

Roof morphology constraints

In the case of the roof morphology, we have as hard constraints the following binary relation for the definition of roof in particular building composition: The relation Covers in order to declare that one roof will cover from one to many spatial objects. With this constraint, a roof inherits the dimensions of a space. In the case of more than one spaces the roof inherits the dimensions of the resulting bounding box that these spaces form. At this stage of our research, we do not utilise any other hard constraint for the roof. In order to obtain the characteristic roof forms, which belong to particular styles, we define a number of new soft constraints. All these constraints control the geometric parameters for the formation of a roof. Their application is for a specific roof each time. However, they should be repeated according to the number of the roofs in a building design. They differentiate in two categories. The first category contains these which control the consistency of the roof morphology, Roof_Homogeneity, South-Side_Homogeneity, North-Side_Homogeneity, East-Side_Homogeneity, West-Side_Homogeneity, Parallel-to_Longer-Dim, Parallel-to_Shorter-Dim, Rectangular_Base, Circular_Base, Roof-Center_in_Center.

The second category contains constraints that control the degree of curvature, linearity, cavity of the roof form. In general, they define three control points of the eight curves that constitute the roof form. Such as, More Cable_than_Vault, More_Vault_than_Cable, More_Flat_than_Cable, More_Flat_than_Vault, More_Vault _than_ Flat, More_ Cable_than_Flat, More_ Pyramidal_than_Flat, ect.

6.3.1.2 Measure of Style

Qualitative aspects are expressed as a set of objective criteria; therefore, we express the degree of adaptation to them as an aggregate of these objective criteria for a 3D scene synthesis [42]. A scale is created, with total adaptation (i.e. a scene is 100% adapted) at one end, and lack of adaptation at the other (i.e. 0% adaptation). The general formal expression of the evaluation for a solution s is given by:

$$SE(s) = g\{f(q_1), f(q_2), \ldots, f(q_n)\} \quad \in \quad [0, 100] \tag{1}$$

where q_i (with $i = 1, 2, \ldots, n$) is the set of n evaluation criteria that express a specific set of qualitative aspects. The $f(q_i)$ expresses the local metrics of each participated evaluation criterion illustrating how each objective criterion in the proposed evaluations contributes to the overall evaluation of a solution for a set of qualitative aspects. The mechanism employed to combine and exploit these partial evaluations to assess the overall solution performance is an intergrading function g in an n-dimensional space, where n depends upon the number of soft constraints that participate in the expression of a set of qualitative aspects. In order to express the set of qualitative aspects of a scene synthesis we imply the calculation of the sum of the number of the stylistic objective criteria. Therefore, the above equation will obtain the form of a linear summation of the weighted evaluations.

$$g\{\} = \sum_{i=1}^{n} a_i f(q_i) \quad 0 \leq a_i \leq 100 \tag{2}$$

For each objective criterion, a constant weighting factor a_i is provided.

6.3.2 Multi-objective Genetic Algorithm

In many synthesis problems, the simultaneous satisfaction of several and in many cases contradictory, objectives is a necessity. Multi-objective optimization is defined as the problem of finding a vector of decision variables, which satisfies constraints and optimizes a vector function whose elements represent the objective functions [18]. A Multi-Objective Genetic Algorithm is introduced into the Declarative Modelling cycle (Figure 6.5.).

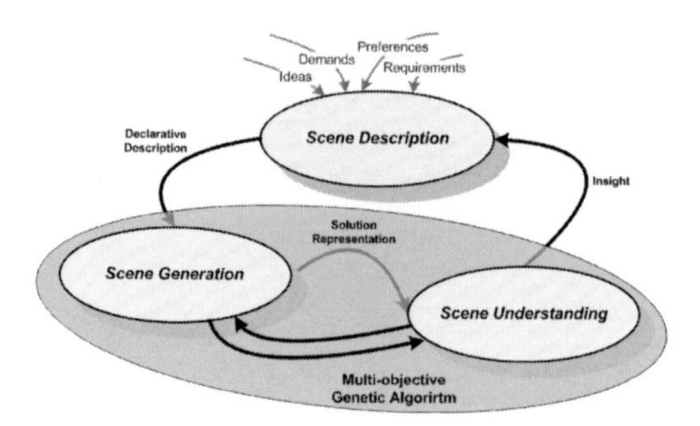

Fig. 6.5. Scope of Multi-Objective GA module

6.3.2.1 Genetic Algorithm

We have developed a system that is based on a Multi-Objective Genetic Algorithm (MOGA) variation based on the Weighted Sum method [35], [44]. In particular, every chromosome is arranged in a hierarchy consisting of multiple pieces of genes, for two fundamental reasons. First, because of the application of declarative modelling by hierarchical decomposition. Secondly, because this arrangement is corresponding to the decomposition – part-of representation of a scene as it is normalized, by the DKABM [42]. Finally, each gene is being defined by float values. The length of a chromosome varies because different number/types of objects can be selected. However it is possible the generation of 3D scenes with varying number of objects. In general, this chromosome may be mapped to a potential solution to the problem being investigated or as the design solution being developed in the case of our research. Based on the hierarchical design representation, the phenotype is represented as the combination of objects and their associated properties. Since every object is defined by its associated object attribute parameters, each chromosome consists of a list of multiples of attribute parameters. The chromosome encoding scheme is based on the real lists, which is proved to be more suitable for this problem solving case. The GA will manipulate the values of these parameters, which they must be coded as the genotype.

An object chromosome is build up from seven genes. The first is decoded into the type of space index, the next triple of genes correspond with positional three-dimensional coordinates of the lower left vertex of the quadrilateral. The next 3 numbers are decoded into the Length, Width and Height of a quadrilateral.

The structural chromosomes stand for the structural objects of a declarative scene description. We consider as a structural object only a Roof. According the geometric description of a Roof the chromosome has 28 genes. In particular, we consider that a Roof chromosome has the typical three genes for position, and three genes about length, width and height. In our case we decide that the attributes of positioning and length-width are defined from the space that a roof covers. The remaining genes that characterise a Roof are D_{OFFSET}_Length, D_{OFFSET}_Width, O_{OFFSET}_Length, and O_{OFFSET}_Width, W_1-W_8, and B_1-B_4.

The problem of determination of the overall relative fitness of the genotypes it has its origins on the multi-objective character of design problems. In order to judging the overall fitness of the solutions, we need an appropriate evaluation mechanism. Therefore, we imply a range-independent multi-objective ranking technique [7]. Such technique enable an equally treatment of the objectives. Moreover it will avoid any laborious fine-tuning of weights. Following this, that technique will facilitate the specification the importance factor of each objective.

6.3.2.2 MOGA Mechanism

The mechanism works as following. This aggregation based mechanism converts the fitness values for each objective into ratios, using the globally best and worst values. These global rations are then summed to provide an overall fitness value for each solution [7]. Once the individuals in the population of the GA have been evaluated, the multi-objective method Sum of Weighted Global Ratios (SWGR) must calculate the overall fitness ranking position of each individual. First, SWGR records global minimum and maximum fitness values for each of the separate fitness values in each individual. For example, if the selected qualitative evaluation objectives calculate five separate fitness values for each individual, then SWGR holds a list of five corresponding minimum and maximum fitness values. These values are updated every generation by examining the fitness values of the new individuals in the internal population. If any fitness value falls below the minimum recorded, then this minimum value is updated. If any value falls above the maximum, then this maximum value is updated. These continuously updated minimum and maximum values, produced so far by a fitness function, give a steadily improving approximation of the effective range of that function. SWGR uses this information to convert every fitness value of every individual into a fitness ratio f. Formally, for any solution s_k, with respect to the qualitative characteristic q:

$$f_k(q_i) = \frac{s_k(q_i) - \min\left(s(q_i)\right)}{\max\left(s(q_i)\right) - \min\left(s(q_i)\right)} \tag{3}$$

Such approach allows the equal treatment of all evaluation criteria. On the other hand it eliminates the difficulty of having multiple criteria with different effective ranges by equalizing all of the effective ranges. Consequently, the individual's ratios are summed, and each fitness value multiplied by its relative importance weight. In this way, a single overall fitness for each individual is achieved. In the MOGA a single point crossover occurs at the bottom of the population structures among the variables that represent the rooms/roofs of a scene. This assures that crossover operations occur only between the same members of the scene, meaning that it is not logical to combine a kitchen's width and a bathroom's length to create a bedroom. We apply random mutation, which helps the MOGA to explore a much greater space on the problem space. In this case for each variable that is going to be mutated, choose a random value within its range and assign this value to the variable. Therefore, every value is possible. The mutation occurs at the end of every evolutional step except the final one, so as not to mutate the resulting population. The mutation procedure defines a random number in the range in which the variable to be mutated is defined. The selection technique is based on fitness proportionate selection. In particular, we imply the "Stochastic Universal Sampling" scheme [44]. In order to preserve individuals corresponding to good solutions we also implement the elitism scheme.

6.4 Implementation Framework

The methodology follows the Declarative Evolutionary Design Cycle and provides the sequence of the phases as implemented. The Model Description phase is broken down to the declarative Model Description and the Style Selection. Both of these activities can be user inputted or make use of stored data. The Generation phase is supported by the generators parameters definition and the solutions generation as well as the generation engine (multi-objective genetic algorithm) fine tuning which is also the feedback supplied by the solution understanding phase. The generation phase leads to the generated solutions, which are data, stored in the Scene base of the MultiCAD Knowledge base. The Solutions Understanding phase consists of the graphical visualization of the generated solutions and the evaluation of the user that can lead to further fine-tuning and solutions generation or to the end of the flowchart. The knowledge development is the phase where the developer (expert) defines both the attributes of DKABM, and architectural styles for use in the style selection by the designer [42].

6.4.1 Software Architecture

The architecture of the current implementation is based on the architecture proposed by the MultiCAD framework. Therefore, the software is divided into three main layers that have specific roles in the implementation. The three layers are the following: the User Interface Layer, the Processing Layer (the solutions generator) and finally the Data Management Layer (Figure 6.6.).

The *User Interface* Layer is the topmost layer and consists of the front end interfaces with which the user interacts. The *Processing* Layer is the middle layer and contains the solutions generator, which receives information from the top layer and creates geometrical solutions. The bottom layer, the *Data Management* Layer, concerns the data and knowledge storing and retrieving. Besides, the layer receives the geometrical solutions from the processing layer and knowledge from the topmost layer via knowledge acquisition.

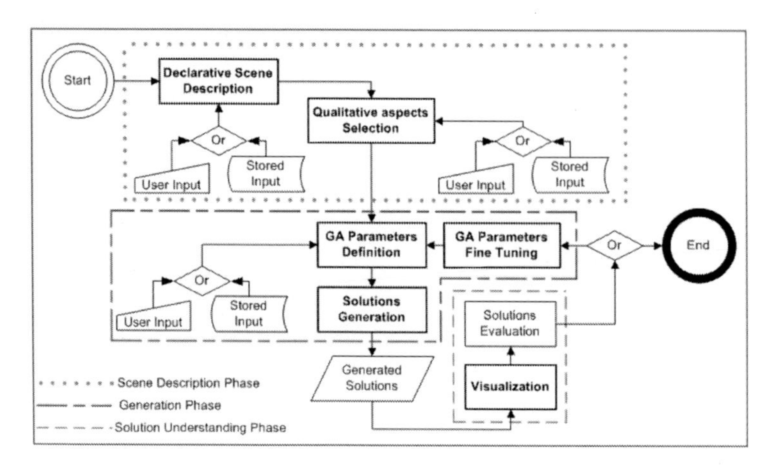

Fig. 6.6. Flowchart of MOGA operation

6.4.1.1 User Interface Layer

The *User Interface* Layer consists of the forms – interfaces that the user has access to. Moreover it is divided into the following five main components (Figure 6.6.).

***Declarative Model Description* component.** The set of forms with which the user presents the model to the system. Here the user defines model information such as number of rooms, room types, site's properties, rooms' properties, relations between rooms and roofs and hard constrains.

***Style Definition* component.** It consists of two forms that are accessed only by the developer (power user) and give him/her the ability to define fitness functions and weights, for rooms and roofs respectively, which define an architectural style.

***Style Selection* component.** This is the equivalent of the Style Definition component for the simple user. In this case the user will be called to select a style among a list of predefined, by the developer, architectural styles.

GA Control. A component where the GA's parameters are defined. Such as number of atoms per population, number of generations, probability of mutation, the atoms' max age, the elitism et cetera, and have great influence over the outcome of the generator. Moreover, the user defines parameters about the number of repetition (loops), the number of atoms per generation to be stored in the Scene Base, the number of generations required to pause the evolution in order to have the ability to visualize partial results.

***Visualization* component.** A component accesses the stored geometrical solutions and presents them to the user graphically.

***Knowledge Development* component.** The knowledge development component is the interface that provides the system with knowledge concerning the DKABM and architectural styles.

6.4.1.2 Processing Layer

The Processing Layer consists of one component that is the GA Geometrical Solutions Generator. This component receives the declarative model description, the style information and the GA properties previously defined in the upper layer and computes geometrical solutions genetically (Figure 6.6.). For the computation of the geometrical solutions, the GA Geometrical Solutions Generator component follows five main steps.

Initial Population. The solutions generator receives and is based on the declarative model description, and builds the initial population. The latter is setup based on the properties defined in the user interface layer and the model constrains.

Fitness Score Calculation. Once the initial population is built the GA computes the fitness scores of the current population based on the fitness functions defined in the user interface layer.

Atoms Selection. The individuals' selection receives the calculated fitness scores and selects individuals accordingly to fill up the matting pool for the evolution of the population.

Crossover. In this stage individuals from the mating pool are selected and combined to build up the evolved atoms in the new population.

Mutation. A possibility of mutation was defined in the previous layer in the GA control component and based on that some individuals will be mutated.

Final Population. The final population is the evolved set of generated solutions. This stage is reached only when the defined, in the GA properties, generations have finished. If there are more generations left then the evolution after the mutation stage will return to the fitness calculation stage for further evolving.

6.4.1.3 Data Management Layer

The final layer is the Data Management Layer that contains the MultiCAD intelligent Base. The intelligent database consists of two main databases that communicate with each other (Figure 6.6.). These bases are the following.

- *Scene Base.* The Scene base consists of multiple data tables that store all the information concerning the scenes, solutions, fitness functions' scores and weights along with generations and repeat times.
- *Knowledge Base.* The Knowledge base contains two bases: the Style base and DKABM base. The Style base contains all the information concerning architectural styles. The DKABM base holds the information concerning DKABM patterns.

6.5 System Evaluation

In this framework according to the DM cycle, we define the following steps. The designer introduces type and number of all spaces of the building. The designer introduces information about the placement of all spaces as a range that is limited by the site limits; it is possible to select some fixed position for one or more spaces within a particular part of the site. Next, he/she defines the range of their length, width and height. It is possible to define either fixed and/or variable dimensions for each space. The designer defines specific requirements for the building as constraints between the spaces. Such relations, in general, they have the form of binary topological relations. In the next step, the designer introduces a set of qualitative aspects that a scene/subscene should achieve. In particular, such aspects are defined in the form of weighted evaluation criteria within a multi-objective function [42]. The designer implements the genetic algorithm with some pre-specified properties, which they could be redefined in order to adapt better the GA to a particular scene synthesis problem, (Figure 6.6.). For adaptation to the designer qualitative aspects the MOGA use a number of fitness values as generated by the evaluation function for each scene, in order to guide evolution towards scene solutions adapted to particular aspects [42]. Designer's set of qualitative aspects are mapped to objectives that require different importance weight value. Qualitative aspects are expressed as a set of objective criteria; therefore, we express the degree of adaptation to them as an aggregate of these objective criteria for a 3D scene synthesis [42].

We perform a series of experiments for system evaluation and validation, which are based on two alternative sets of qualitative aspects. The latter have been defined to

Table 6.1. Designers' views from Santorini and Metsovo

Santorini aspects	Metsovo aspects
Bedroom and living room are coaxial	Bedroom and living room are coaxial
Bedroom width is equal Living room width	Building plan is longer than wide (deep)
Building plan is wider (deeper) than long	Building plan is rectangular
Rooms are deeper than long	Rooms are deeper than long
Rooms have quadrilateral form	Rooms have quadrilateral form

Table 6.2. Santorini and Metsovo aspects are expressed as a set of weighted objective criteria

Santorini objective criteria		Metsovo objective criteria	
Local	Weight	Local	Weight
Deeper than Long	1	Longer than Deep	1
Same Width	1	Deeper than Long	1
		Same Length	1
Global		Global	
Building Compactness	10	Building Compactness	10
Non Overlapping	10	Non Overlapping	10
Building Deeper than Long	5	Building Longer than Deep	5

reflect local designers' views [43] originating from Santorini and Metsovo respectively, (Table 6.1.). In this way, the plan of the experiment is organised as following.

The Santorini and Metsovo intent is expressed as a set of weighted objective criteria, (Table 6.2.). The evolutionary process concerns the adaptation of a scene to criteria of spatial planning and roof morphology.

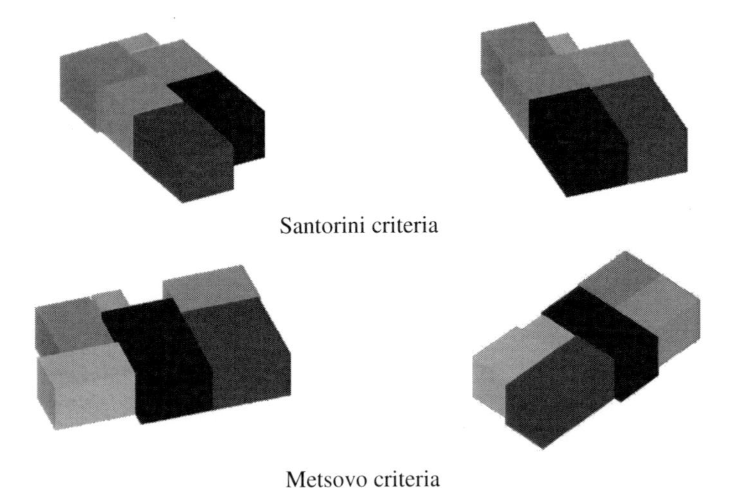

Fig. 6.7. Generated adapted scenes

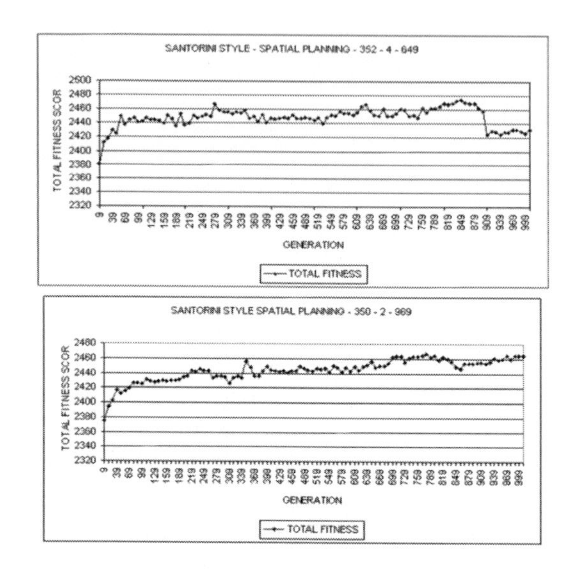

Fig. 6.8. Process of evolution

From the current experiment, well adaptive scene synthesis was evolved (Figure 6.7., 6.9.), besides, they did not violate hard constraints that imposed by the designer's description.

However, all objective criteria are optimized at an appropriate level (Figure 6.8.), given the fact of their relevant degree of importance. The evolved synthesis were along the evolution obtain the special characteristics of either Santorini or Metsovo set of qualitative aspects concerning both scene's spatial planning (Figure 6.7., 6.9.). Solutions are similar to original examples of edifices from each set of qualitative aspects. From observations of the best synthesis scores, we saw that the MOGA converge early to accepted solutions (approximately beyond 450th-650th of 1000 generations). The computing time was characteristically short.

In order to evaluate the performance of the system we provide a framework for the experiments. The performance of the system is an important criterion for the feasibility of the system. In this experiment we try to evaluate the performance of the Evolutionary Declarative design system prototype when it is confronting simple building designs with no more than six spaces and at the same time with few hard constraints among them. We enable the spaces to have varied dimensions while they are only constrained by the building brief. A number objective functions as expression of stylistic principles direct the evolution of the individuals in order for the later to express building designs adapted to an architectural style.

The whole process has two distinct steps. Firstly, the designer utilise the MOGA system in order to evolve the spatial planning – composition for given building requirements, under specific principles of an architectural style. Secondly, from the results of the earlier step, in particular from the resulted individual of the last generation, the designer selects one individual in order to evolve the roof morphology of the building. The evaluation criteria are specific principles of an architectural style. For the purpose of our work, we examine two different architectural styles, Santorini, and

Santorini criteria (roof style)

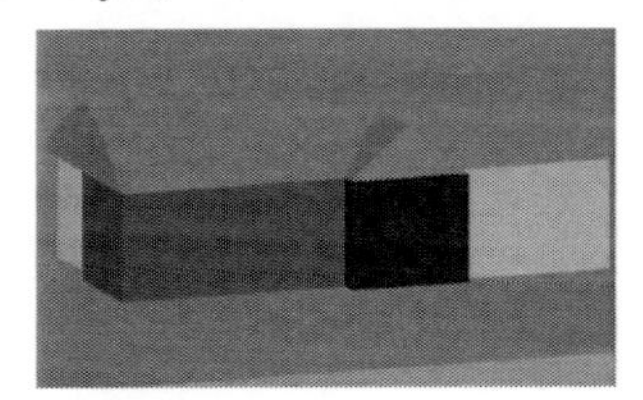

Metsovo criteria (roof style)

Fig. 6.9. Generated adapted 3D scenes

Metsovo. They are quite different on both their spatial composition and roof morphology aspects. Both styles were presented in details in an earlier chapter. In the current experiment, we set up two design case studies and we introduced it within the evolutionary declarative design system. For the firs study a set of evaluation criteria was based on the design principles of architectural style of Santorini. For the second study a set of evaluation criteria based on Metsovo style. From a series of consecutive runs of the system under the description of two alternative settings for the MOGA, for the same Scene we obtain the following results.

The Evolutionary Design system has produced well adaptive building designs according to the stylistic criteria for both styles. Besides, the solutions were without violation of the hard constraints that imposed by the designer description. The MOGA method has dealt with seven objective evaluation functions in the case of Santorini, and ten objectives in the case of Metsovo style. We observe that all objective criteria optimised at an appropriate level, given the fact of their relevant degree of importance. The evolved designs were along the evolution obtain the special characteristics of either Santorini (Figure 6.7., 6.9.) or Metsovo style (Figure 6.7., 6.9.) concerning both the building's spatial planning and the building's roof morphology. Along the consecutive runs of the evolutionary design environment the system provided with well performance, while the MOGA avoids get trapped to local optima. The MOGA has provided the designer of the system with successful designs while it provides variations of the specific architectural style. Especially the later result proves that the selection of the specific stylistic criteria were the appropriate in order to provide the designer with feasible and alternative variations of the same style.

Furthermore, the charts remind us that the scope of the MOGA it is not to obtain the optimum solution. The scope it is rather to provide a range of high score solutions well adapted to the objectives of the particular style. Another result was that the specific evolutionary design environment based on MOGA could enable the easy use of stylistic objective criteria.

The MOGA has provided the designer of the system with successful designs while it provides variations of the basic roof morphology of the two specific architectural styles. Especially the later result proves again that the selection of the specific stylistic criteria were the appropriate in order to provide the designer with feasible and alternative variations for roof morphology of the same style. By the generated solutions is proved that the selection of the specific stylistic criteria were the appropriate in order to provide the designer with feasible and alternative variations of the same style. Another outcome was that the specific evolutionary design environment based on genetic algorithm could enable the easy use stylistic objective criteria. We consider three aspects of particular interest when assessing the performance of the evolutionary methods: effectiveness, efficiency and robustness. Effectiveness usually refers to the quality of the solutions produced by the method. Efficiency usually refers to how much computation time and memory the method uses. Robustness usually refers to how consistent the method is in producing the same or very similar results over many runs on the same problem instance. The quality of the solutions is observed as very high based on the fact that the generated solutions are similar to original examples of edifices from each style. As it concerns the efficiency aspect of the system, we observe that the system is consume very little computation time and memory. Furthermore, the applied MOGA method provides high degree of robustness. In all runs, the method generates well-performed solutions. Additionally, it provides the designer with variation within the specific architectural style.

6.6 Discussion

In order to take advantage of the capabilities of declarative modelling in architectural design, we confronted the problems from two points of departure considering the three phases of declarative conceptual cycle.

First, up to now declarative modellers have use constraint satisfaction techniques for the exploration of solution space and consequently the generation of solutions. These approaches are interesting for an exhaustive search of the solutions space. Along with drawbacks and limitations (long generation time and numerous scene solutions) there were presented in many cases successfully approaches. However these approaches are not very appropriate for architectural design. The evaluation of the alternative solutions it generally concerns numerous criteria, in many cases complex and conflicting among each other. In many architectural design problems, the simultaneous satisfaction of several objectives is essential. An architect during conceptual design needs to search huge combinatorial spaces in a non-exhaustive and guided way Therefore we propose a new type of generative engine, that of genetic algorithm. The implementation of a multi-objective genetic algorithm (MOGA) provides promising results. Furthermore, during the scene understanding phase the designer has a tedious and hard task to complete because of the numerous resulted scenes. The introduction of multi-objective generates set of solutions already evaluated following specific criteria (soft constraints). For this reason the designer has a significantly less number of solutions to evaluate.

Second, usually the declarative modelling has dealt with hard constraints, which all of them have to be satisfied. Conceptual design determines the principle of a solution. A number of basic problems of conceptual design are the following. In a given problem, there exist constraints and objectives but their distinction and classification is very often very fuzzy. In many cases examination and understanding of the problem ends up with a move from objectives to constraints or vice versa. The nature of constraints could be either hard or soft. The continuous expanding of problem demands some of them could change character from hard to soft and vice versa. It is ever possible the elimination of some of constraints during the process. However, in conceptual architectural design there are hard constraints and soft constraints. In order to exploit the capabilities of declarative modelling in architectural design we provide such separation of constraints. Then we introduce soft constraints as objective functions in a multi-objective genetic algorithm. In a series of experiments, we observe that the MOGA constantly did not violate hard constraints, while it generates results well adapted to soft constraints.

6.7 Conclusions

The main hypothesis of the project, if it is possible the introduction of the dimension of style in the frame of declarative modelling in order to aid architectural design during the conceptual phase, has a positive answer.

A methodology of architectural conceptual design utilising style, and based in the frame of declarative modelling, and evolutionary algorithms has a significant result. The result is the system prototype for Evolutionary Declarative DesigN -MultiCAD (EDDEN-MultiCAD) for the aid of conceptual architectural design. The satisfaction of our hypothesis beyond the apparent result of the prototype system is beneficial to a wide range of design-related phenomena.

The resulted prototype system supporting building synthesis takes into consideration aesthetic intent and particular the dimension of style. The designer is able to design buildings adapted to a particular architectural style.

The resulted method could successfully quantifying aesthetic qualitative criteria of a building. A model of style could enable the representation of style. The quantification of style could provide architects with more decision criteria during the generation of solutions. Architects have an operational tool to model his/her aesthetic intent, and further evolve them. In general, such tools could provide the ability to designers to distil complex and qualitative criteria. Moreover, it is a promising Design Decision Support system based on aesthetic criteria. It is evident that such a system points towards the direction of a Computer Aided Aesthetic Design tool.

6.7.1 Declarative Modelling and Architectural Conceptual Design

The declarative conceptual cycle provide an efficient tool for architectural conceptual design. The introduction of architectural knowledge and style in parallel with the adoption of evolutionary algorithms improve its capabilities. The applicability of declarative modelling systems in architectural design is great. The introduction of evolutionary declarative design paradigm in Computer Aided Architectural Design addresses the demanding early phase of architectural design process. Such a tool will increase the designer's ability in generating novel design concepts.

6.7.2 Aesthetic and Artificial Intelligence

The aesthetic dimension in the form of stylistic principles can be supported by artificial intelligence techniques. Sets of stylistic criteria (aesthetic intent) have been used for both generation and evaluation of design products (building) and aids the designer's final decisions. Such methodology amplifies creativity in the design process. Architects and researchers are provided with a tool to achieve a quantitative understanding of the aesthetic issues involved in the design process.

The approached method could have an influential impact on studies about style formation, representation and use in architectural design and beyond. Moreover, it could be a promising tool for the studies on building morphology. Morphology, the study of pattern and form, is crucial to design because it constitutes an essential part of its corpus of coherent knowledge.

In general the present application can make a contribution to architectural research and design, and further could have a feedback on design theory.

However, the design environment for the implementation of our ideas was the conceptual design of buildings, many methods and techniques could extend towards other areas of design. The Evolutionary Declarative Design-MultiCAD system with the appropriate adaptations could be applicable to a wider type of product's conceptual design, like interior design, furniture design and industrial design.

References

[1] Ackerman, J.S.: Style. In: Ackerman, J.S., Carpenter, R. (eds.) Art and Archaeology, pp. 174–186. Prentice-Hall, Englewood Cliffs (1963)

[2] Akin, O.: Psychology of Architectural Design. Pion London (1986)

[3] Banzhaf, W., Nordin, P., Keller, R., Francone, F.: Genetic Programming An introduction. Morgan Kaufmann, San Francisco (1998)

[4] Bardis, G.: Machine Learning and Decision Support for Declarative Scene Modelling / Ap-prentissage et aide à la décision pour la modélisation déclarative de scènes (bilingual), Thèse de Doctorat, Université de Limoges, France (2004) (2006)

[5] Bense, M.: Aesthetica. Einfuehrung in die neue Aesthetik. Agis-Verlag, Baden-Baden (1965)

[6] Bentley, P. (ed.): Evolutionary Design by Computers. Morgan Kaufmann Publishers, San Francisco (1999)

[7] Bentley, P.: Generic Evolutionary Design of Solid Objects Using a Genetic Algorithm PhD thesis University of Huddersfield (1996)

[8] Blrkhoff, G.D.: Aesthetic Measure. Harvard University Press, Massachusetts (1933)

[9] Bonnefoi, P.F., Plemenos, D.: Constraint satisfaction techniques for declarative scene modelling by hierarchical decomposition. In: 3IA 2000 International Conference, Limoges (2000)

[10] Bremen van, E.J.J., Sudijone, S., Horvath, I.: A contribution to finding the relationship between Shape Characteristics and Aesthetic Appreciation of Selected Product. In: Lindemann, U., Birkhofer, H., Meerkamm, H., Vajna, S. (eds.) Proc. 12th International Conference on Engineering Design ICED 1999, pp. 1765–1176 (1999)

[11] Caldas, L.G.: An Evolution-Based Generative Design System: Using Adaptation to Shape Architectural Form. PhD thesis Massachusetts Institute of Technology (2001)

[12] Chan, C.S.: Exploring individual style in Design. Environment and Planning B: Planning and Design 19, 503–523 (1992)

[13] Chan, C.S.: Operational definition of Style. Environment and Planning B: Planning and Design 21, 223–246 (1994)

[14] Chen, K., Owen, C.: Form language and style description. Design Studies 18, 249–274 (1997)

[15] Chiou, S.C., Krishnamurti, R.: The grammar of Taiwanese traditional vernacular dwellings. Environment and Planning B: Planning and Design 22, 689–720 (1995)

[16] Coates, P., Broughton, T., Jackson, H.: Exploring Three-Dimensional Design Worlds using Lindenmayer Systems and Genetic Programming. In: Bentley, P.J. (ed.) Evolutionary Design by Computers. Morgan Kaufmann, San Francisco (1999)

[17] Coates, P., Makris, D.: Genetic Programming and Spatial Morphogenesis. In: AISB 1999 Symposium on Creative Evolutionary Systems Edinburgh (1999)

[18] Coello, C.A.C.: A Comprehensive Survey of Evolutionary-Based Multiobjective Optimization Techniques. Knowledge and Information Systems 1(3), 269–308 (1999)

[19] Cvetkovic, D.: Evolutionary Multi-Objective Decision Support Systems for Conceptual Design. PhD Thesis School of Computing University of Plymouth (2000)

[20] Ding, L., Gero, J.: The emergence of the representation of style in design. Environment and Planning B: Planning and Design 28, 707–731 (2001)

[21] Dragonas, J.: Collaborative Declarative Modelling / Modelisation Declarative Collaborative (bilingual), Thèse de Doctorat, Université de Limoges, France (2006)

[22] Fogel, L.J., Owens, A.J., Walsh, M.J.: Artificial Intelligence through Simulated Evolution. Wiley, New York (1966)

[23] Frazer, J., Tang, M.X., Sun, J.: Towards a Generative System for Intelligent Design support. In: Proceedings of the 4th CAADRIA Conference (1999)

[24] Frazer, J.H.: An Evolutionary Architecture. Architectural Association Pubs., London (1995)

[25] Fribault, P.: Modélisation Déclarative d'Espaces Habitable (in French). Thèse de Doctorat Université de Limoges, France (2003)

[26] Furuta, H., Maeda, K., Watanabe, E.: Application of Genetic Algorithm to Aesthetic Design of Bridge Structures. Microcomputers in Civil Engineering 10(6), 415–421 (1995)

[27] Gero, J., Kazakov, V.: Adaptive enlargement of state spaces in evolutionary designing. Artificial Intelligence in Engineering Design, and Manufacturing 14, 31–38 (2000)

[28] Giannini, F., Monti, M.: Design intent-oriented modelling tools for aesthetic design. In: WSCG 2003 conference Plzen, Czech Republic (2003)

[29] Giannini, F., Monti, M., Podehl, G.: Aesthetic-driven tools for industrial design. Journal of Engineering Design 17(3), 193–215 (2006)

[30] Goldberg, D.: Genetic Algorithms in Search, Optimization, and Machine Learning. Addison-Wesley, Reading (1989)

[31] Golfinopoulos, V.: Etude et Réalisation d'un Système de Retro-Conception base sur la Connaissance pour la Modélisation Déclarative de Scènes. PhD thesis (In english and french), Université de Limoges, France (2006)

[32] Gombrich, E.H.: Art and Illusion. Pantheon, New York (1960)

[33] Grierson, D.E.: Conceptual Design Using a Genetic Algorithm. In: Proceedings of the 15th Structures Congress "Building to Last", Part 2 (of 2), pp. 798–802 (1997)

[34] Harrington, S.J., Naveda, J.F., Jones, R.P., Roetling, P., Thakkar, N.: Aesthetic Measures for Automated Document Layout. In: Doceng 2004 (2004)

[35] Holland, J.H.: Adaptation in Natural and Artificial Systems. University of Michigan Press (1975)

[36] Jo, J.H., Gero, J.S.: Space Layout Planning Using an Evolutionary Approach. Artificial Intelligence in Engineering 3(12), 149–162 (1998)
[37] Jupp, J., Gero, J.S.: Visual Style: Qualitative and Context Dependent Categorization. Artificial Intelligence for Engineering Design, Analysis and Manufacturing 20, 247–266 (2006)
[38] Kanal, L., Kumar, V. (eds.): Search in Artificial Intelligence. Springer, Heidelberg (1988)
[39] Koile, K.: Formalizing Abstract Characteristics of Style. Artificial Intelligence for Engineering Design. Analysis and Manufacturing 20, 267–285 (2006)
[40] Koza, J.R.: Genetic Programming. MIT Press, Cambridge (1992)
[41] Lavie, T., Tractinsky, N.: Assessing dimensions of perceived visual aesthetics of web sites. Int. J. Human-Computer Studies 60, 269–298 (2004)
[42] Makris, D.: Study and Realisation of a Declarative System for Modelling and Generation of Style with Genetic Algorithms. Application in Architectural Design / Etude et réalisation d'un système déclaratif de modélisation et de génération de styles par algorithmes génétiques. Application à la création architecturale (bilingual). Thèse de Doctorat, Université de Limoges, France (2005)
[43] Michelis, P.: Aesthetic studies. vol. 1-2 P. E. Michelis Foundation, Athens, Greece (2001)
[44] Mitchell, M.: An introduction to Genetic Algorithms. MIT Press, Cambridge (1998)
[45] Ngo, D.C.L., Teo, L.S., Byrne, J.G.: Evaluating Interface Esthetics. Knowledge and Information Systems 4, 46–79 (2002)
[46] Oduguwa, V., Tiwari, A., Roy, R.: Evolutionary computing in manufacturing industry: an overview of recent applications. Applied Soft Computing 5, 281–299 (2005)
[47] O'Reilly, U.-M., Ramachandran, G.: A preliminary investigation of evolution as a form design strategy. In: Adami, C., Belew, R., Kitano, H., Taylor, C. (eds.) Proceedings of the 6th conference on Artificial Life. MIT Press, Cambridge (1998)
[48] Orsborn, S., Cagan, J., Pawlicki, R., Smith, R.C.: Creating Cross-Over Vehicles: Defining and Combining Vehicle Classes Using Shape Grammars. Artificial Intelligence for Engineering Design, Analysis and Manufacturing 20, 217–246 (2006)
[49] Park, H.-J.: A Quantification of Proportionality Aesthetics in Morphological Design Ph. D. thesis University of Michigan (2005)
[50] Park, K.W., Grierson, D.E.: Pareto-optimal Conceptual Design of the Structural Layout of Buildings Using a Multicriteria Genetic Algorithm. Journal of Computer-Aided Civil and Infrastructure Engineering (1998)
[51] Plemenos, D.: Contribution à l'étude et au développement des techniques de modélisation, génération et visualisation de scènes – Le projet MultiFormes. PhD dissertation (1991)
[52] Plemenos, D.: Declarative Modeling by Hierarchical Decomposition. The actual state of the MultiFormes Project. In: GraphiCon 1995, St. Petersbourg, Russia (1995)
[53] Podehl, G.: Terms and Measures for Styling Properties. In: Proceedings of the 7th International Design Conference Dubrovnik, Croatia, Zagreb, pp. 879–886 (2002)
[54] Prats, M., Earl, C., Garner, S., Jowers, I.: Shape Exploration of Designs in a Style: Toward Generation of Product Designs. Artificial Intelligence for Engineering Design, Analysis and Manufacturing 20, 201–215 (2006)
[55] Renner, G., Ekart, A.: Genetic algorithms in computer aided design. Computer-Aided Design 8(35), 709–726 (2003)
[56] Rowe, P.: Design Thinking. MIT Press, Cambridge (1991)

[57] Sanchez, S., le Roux, O., Luga, H., Gaildrat, V.: Constraint-Based 3D-Object Layout using a Genetic Algorithm. In: Plemenos, D. (ed.) Proc. 3ia 2003 - 6th International Conference on Computer Graphics and Artificial Intelligence Limoges, France (2003)

[58] Schwefel, H.-P.: Numerical Optimisation of Computer Models. Wiley, Chichester (1981)

[59] Sequin, C.: CAD tools for aesthetic engineering. Computer-Aided Design 7(37), 737–750 (2005)

[60] Shapiro, M.: Style. In: Philipson, M. (ed.) Aesthetics Today, pp. 81–113. Word Publishing (1961)

[61] Shrestha, S.M., Ghaboussi, J.: Evolution of Optimum Structural Shapes Using Genetic Algorithm. ASCE Journal of Structural Engineering 11(124), 1331–1338 (1998)

[62] Simon, H.A.: Style in design. In: Eastman, C. (ed.) Spatial Synthesis in Computer-aided Building Design Applied Science, London, pp. 287–309 (1975)

[63] Smithies, K.W.: Principles of Design in Architecture. Van Nostrand Reinhold, New York (1981)

[64] Staudek, T.: Computer-aided aesthetic evaluation of visual patterns. In: Proceedings Conference ISAMA-BRIDGES, pp. 143–149 (2003)

[65] Stiny, G., Grips, J.: Algorithmic Aesthetics. University of California Press (1978)

[66] Sun, J., Frazer, J.H., Mingxi, T.: Shape optimisation using evolutionary techniques in product design. Computers & Industrial Engineering 53, 200–205 (2007)

[67] Testa, P., O'Reilly, U., Greenwold, S., Hemberg, M.: AGENCY GP: Agent-Based Genetic Programming for Spatial Exploration. In: Generative Evolutionary Computer Conference (2001)

[68] Testa, P., O'Reilly, U.M., Kangas, M., Kilian, A.: MoSS: Morphogenetic Surface Structure - A Software Tool for Design Exploration. In: Proceedings of Greenwich 2000: Digital Creativity Symposium (2000)

[69] Vassilas, N., Miaoulis, G., Chronopoulos, D., Constandinidis, E., Ravani, I., Makris, D., Plemenos, D.: MultiCAD-GA: A System for the Design of 3D Forms Based on Genetic Algorithms and Human Evaluation. In: Vlahavas, I., Spyropoulos, C. (eds.) Lecture Notes in Artificial Intelligence: Methods and Applications of Artificial Intelligence, Second Hellenic Conference on AI, Thessaloniki, Greece, pp. 203–214. Springer, Heidelberg (2002)

[70] Wang, C., Vergeest, J.S.M., Wiegers, T., van der Pant, D.J., van den Berg, T.M.C.: Exploring the Influence of Feature Geometry on Design Style. In: Mastorakis, N.E., et al. (eds.) Electrical and Computer Engineering Series: Computational Methods in Circuits and Systems Application, pp. 21–28. WSEAS Press (2003)

[71] Wannarumon, S., Bohez, E.L.J.: A New Aesthetic Evolutionary Approach for Jewelry Design. Computer-Aided Design & Applications 1-4(3), 385–394 (2006)

7

Network Security Surveillance Aid Using Intelligent Visualization for Knowledge Extraction and Decision Making

Ioannis Xydas

Department of Computer Science, Technological Educational Institution of Athens,
28 Ag. Spyridonos Str., 12210 Egaleo, Athens, Greece
yxydas@teiath.gr

Abstract. Web sites are likely to be regularly scanned and attacked by both automated and manual means. Intrusion Detection Systems (IDS) assist security analysts by automatically identifying potential attacks from network activity and produce alerts describing the details of these intrusions. However, IDS have problems, such as false positives, operational issues in high-speed environments and the difficulty of detecting unknown threats. Much of ID research has focused on improving the accuracy and operation of IDSs but surprisingly there has been very little research into supporting the security analysts' intrusion detection tasks. Lately, security analysts face an increasing workload as their networks expand and attacks become more frequent. In this chapter we describe an ongoing surveillance prototype system which offers a visual aid to the web security analyst by monitoring and exploring 3D graphs. The system offers a visual surveillance of the network activity on a web server for both normal and anomalous or malicious activity. Colours are used on the 3D graphics to indicate different categories of web attacks and the analyst has the ability to navigate into the web requests, of either normal or malicious traffic. The combination of interactive visualization and machine Intelligence facilitates the detection of flaws and intrusions in network security, the discovery of unknown threats and helps the analytical reasoning and the decision making process.

Keywords: Visual Analytics, Web Visualization, Web Intrusion Detection, Evolutionary Artificial Neural Networks, Network Security, Surveillance Aid.

7.1 Introduction

With the rapid growth of interest in the Internet, network security has become a major concern to companies and organizations throughout the world. The fact that the information and tools needed to penetrate the security of corporate networks are widely available has increased that concern. Because of this increased focus on network security, network administrators often spend more effort protecting their networks than on actual network setup and administration. Tools that probe for system vulnerabilities and some of the newly available scanning and intrusion detection packages and appliances, assist in these efforts, but these tools only point out areas of weakness and may not provide a means to protect networks from all possible attacks. Thus, a network administrator must constantly try to keep abreast of the large number of security issues confronting him into day's world.

G. Miaoulis and D. Plemenos (Eds.): Intel. Scene Mod. Information Systems, SCI 181, pp. 185–214.
springerlink.com

7.1.1 Web Security

In this project we will focus on Web security, as the World Wide Web is the most widespread Internet service today. With the recent explosion of the Internet, e-commerce, and web-based applications, an Internet presence is now essential for companies and organizations. Consumers have come to expect and in some cases demand, the ability to interact with an organization via the web. As a result of this trend, many organizations are scrambling to deploy not only static content to the web, but also feature-rich applications that allow users to purchase goods and services, interact with customer support, manage their accounts and perform many other functions.

However, many times security and development "best practices" take a back seat to ease-of-use and access-speed to a market. In addition, most systems administrators rarely have an opportunity to interact with development teams as applications are being developed. As systems administrators, one of their primary duties is maintaining the integrity and security of their systems and networks. However, even the most impregnable of systems can be quickly compromised by exploiting an insecure application that is running on it. Nowhere is this more evident than on the web.

The threat profile facing organizations has undeniably shifted from network-layer exploits to more formidable attacks against applications, primarily Web and Web services applications.

According to a recent report published by the Common Vulnerabilities and Exposures (CVE) project [1], flaws in Web software are among the most reported security issues so far this year. It is easy to see why. Hackers are known to search for an easy target. Poorly configured or poorly written web applications are not only an easy target, taking the attackers straight to their goal, giving them access to data and other information, but also can be used to spread malicious software (malware) such as viruses, worms, Trojan horses and spyware to anyone who visits the compromised site.

"Easy to learn" scripting languages enable anyone with an eye for graphic design, not necessary with developer background, to develop and code powerful web-based applications. Unfortunately, many developers only bother to learn the eye-catching features of a language and not the security issues that need to be addressed. Also, many of the introductory books on coding fail to discuss security. As a result, many of the same vulnerabilities that were problematic for developers several years ago remain a problem today. This is perhaps why Cross-Site Scripting (XSS) is now the most common type of application layer attack, while buffer overflow vulnerability, the perpetual No. 1, has dropped to fourth place. Two other web application vulnerabilities, SQL injections and PHP remote file inclusions, are ranked second and third today [2].

7.1.2 Intrusion Detection

The work of an Intrusion Detection (ID) analyst is a complex task that requires experience and knowledge. Intrusion Detection systems (IDS) are an indispensable part of the information security infrastructure of every networking company or organization. Analysts must continually monitor Intrusion Detection Systems (IDS) for malicious activity. The number of alerts generated by most IDS can quickly become overwhelming and thus the analyst is overloaded with information which is difficult to monitor and analyze. Attacks are likely to generate multiple related alerts. Current IDS do not make it easy for operators to logically group related alerts. This forces the analyst to look only at aggregated summaries of alerts or to reduce the IDS signature set in order to reduce the number of alerts. In the current version of Snort [3], an open source IDS available to the general public, there are more than 12000 signatures for network intrusion detection, over 2300 of which are web-related signatures. By reducing the signature set the analyst knows that although it reduces the false alarms it is also likely to increase the number of false negatives, meaning that he will not be able to detect actual attacks.

Security analysts monitor network activity using an IDS for evidence of actions that attempt to compromise the integrity, confidentiality or availability of a network or computer resource. They use that IDS output in conjunction with other system, network and firewall logs to keep abreast of system activity and potential attacks.

Protecting an application from attack requires a complete understanding of all application communications. Unless a device can "see" the same data as the application it is protecting, it will be unable to identify application-layer threats. This means that to secure any common Web-based application, a security device must perform a full deconstruction of the HTML (Hyper Text Markup Language) data payload, as well as track the state of each application session.

To detect web-based attacks, intrusion detection systems (IDS) are configured with a number of signatures that support the detection of known attacks. Unfortunately, it is hard to keep intrusion detection signature sets updated with respect to the large numbers of continuously discovered vulnerabilities. Developing ad hoc signatures to detect new web attacks is a time-intensive and error-prone activity that requires substantial security expertise. To overcome these issues, misuse detection systems should be complemented by anomaly detection systems, which support the detection of new attacks. Unfortunately, there are no available anomaly detection systems tailored to detect attacks against web servers and web-based applications.

According to a survey [4], in the intrusion detection area visualization tools are needed to offload the monitoring tasks, so that anomalies can be easily flagged for analysis and immediate response by the security analyst. Information presented in a visual format is learned and remembered better than information presented textually or verbally. The human brain is structured so that visual processing occurs rapidly and simultaneously. Given a complicated visual scene humans can immediately pick out important features in a matter of milliseconds. Humans are limited in terms of attention and memory but they excel at the processing of visual information.

7.1.3 Visualization

There are innumerable number of web analysis packages available, commercial or research systems, that present information about the content, structure and usage of web sites. Web analyzers come in all shapes and sizes. Some are better at representing structure, whereas others are more optimized for looking at the content. Commercial offerings help manage large web sites by providing graphic navigation, analysis techniques and conceptual navigation through data. With web security and intrusion detection however there is a lack of visualization tools for monitoring and analysis activities.

The most important source of information for security analysts is the output of IDS, which automatically identify potential attacks and produce descriptive alerts. Due to the complicated nature of detecting actual intrusions, most current IDS place the burden of distinguishing an actual attack from a large set of false alarms on the security analyst, resulting in a significant cognitive load.

This cognitive load born by the security analyst may be mitigated using Information Visualization (IV). Visualization combined with Artificial Intelligence (AI) will take advantage of human perceptual abilities and expertise to amplify cognition.

Although Information Visualization seems like a natural choice for Intrusion Detection, until recently there has been little research into coupling the two technologies.

7.1.4 Visual Data Analysis

Visualization tools and techniques are currently rather weak with regard to large data volumes and complex structures. Besides the imperfection of the existing tools there is a more fundamental reason for this. In visual exploration and analysis, it is the mind of a human explorer that is the primary tool of analysis. It is the task of the human mind to derive insights, "detect the expected and discover the unexpected" while the task of visualization is, defined as, "to make information perceptible to the mind or imagination". However, the human mind has natural limitations as to the amount of information that can be effectively perceived. Therefore, it is often impossible to visualize all data that needs to be analyzed in such a way that the analyst can perceive them all without substantial losses [5].

Visual analytics is the science of analytical reasoning facilitated by interactive visual interfaces [6]. Visual analytics takes advantage of human perception capabilities and can be described as the ability to "find patterns in known and unknown large dataset through visual interaction and thinking". Several new trends are emerging from visual analytics and among the most important one is the fusion of visualization techniques with other areas such as cognitive and perceptual sciences, statistical analysis, mathematics, knowledge representation and data mining.

The basic idea of visual analytics is to visually represent information, allowing the human to directly interact with it, to gain insight, to draw conclusions and to ultimately make better decisions. Visual representation of the information reduces complex cognitive work needed to perform certain tasks. People use visual analytics tools and techniques to synthesize information and derive insight from massive, dynamic and often conflicting data by providing timely, defensible and understandable assessments.

Visual analytics mantra

Unlike the information seeking mantra "overview first, zoom/filter, details on demand" the visual analytics process comprises the application of automatic analysis methods before and after the interactive visual representation is used. This is primarily due to the fact that current and especially future data sets are complex on the one hand and too large to be visualized in a straightforward manner on the other hand. Therefore, the visual analytics mantra is presented [7] as:

"Analyze First
Show the Important

Zoom, Filter and Analyze Further
Details on Demand".

7.1.5 Research Objectives

The fundamental question in this project was the following: Can the visual analytics mantra be applied in the context of data analysis for network security? Can we obtain an "intelligent" visual representation of web attacks and extract knowledge from a network operation graph?

To answer to this question we had to first find a way to capture and analyze the raw data in order to distinguish the web attacks from the raw data, such as the normal web requests. Then we had to find a way to distinguish the different types of web attacks and finally visualize the interesting data and show the important parts to the security analyst.

To achieve this goal we had to design and implement a prototype system. This system was to be a surveillance aid for the web security analyst, providing him with an intelligent visual tool to detect anomalies in web requests by exploring 3D graphs.

Visualizing the raw data is in general unfeasible and rarely reveals any insights. Therefore, the data is first analyzed (i.e. intrusion detection analysis) and then displayed. The analyst proceeds by choosing a small suspicious subset of the recorded intrusion incidents by applying filters and zoom operations. Finally, this subset is used for a more careful analysis. Insight is gained in the course of the whole visual analytics process. In the prototype system we chose to visualize the most important attributes of the raw data as well, the reasons for which will be explained later. Raw data was to be captured online from network traffic but the system was to have the option to treat web logs data as well.

The objective of this project was not to advance the techniques used in the Intrusion Detection area in terms of results. The system improves the analyst response in case of an intrusion by helping him in the decision making. The project has demonstrated that "intelligent" visualization considerably reduces the time required for data analysis and at the same time provides insights which might otherwise be missed during textual analysis. Intelligent visualization offers a powerful means of analysis that can help the security analyst uncover hacker trends or strategies that are likely to be missed with other non-visual methods. Visualization allows him to audit the analytical process, since the operator is examining the web traffic directly and online and is making iterative decisions about what is being presented.

The project was composed of the following phases:

Phase1: Register all known web attacks up to date

Web attacks cover two major fields: Web server and Web applications. The two most popular web servers are Microsoft Internet Information Services (IIS) and the opensource Apache Web server. Both servers, due to their popularity dominate the web server area, although many other servers do exist.

Web server attacks are based on a great number of vulnerabilities of server's software modules, such as Active Server Pages (ASP), Microsoft Data Access Components (MDAC), Remote Data Service (RDS), Internet Explorer (IE) and many others.

Web application layer exploits can be classified as either:

1. Attacks attempting to compromise the integrity or availability of application resources, or
2. Attacks aiming to compromise the trust relationship between an application user and the application.

Phase 2: Grouping of web attacks in classes

Due to the large number of available web attacks an automated classification method should be developed to classify the web attacks and create groups or classes of similar type attacks. To achieve this artificial intelligence was used and specifically Self-Organising Neural Networks technology.

Phase 3: Detection and classification of the web attacks

This phase was the most important and covered the detection of a web attack using automated means. It involved a task to create a knowledge data base using artificial intelligence and machine learning. Artificial Neural Networks (ANN) were primarily used to train the machine to recognize the different kind of web attacks. The system looks into web requests to detect "fingerprints" which are special characters or chains of characters. These fingerprints are then passed to an expert system to decide if they constitute a malicious request or attack.

Due to some drawbacks of Neural Networks, discovered after the first implementation of the prototype system, a hybrid expert system was used as the knowledge data base. It is an Evolutionary Artificial Neural Networks combining Neural Networks and Genetic Algorithms for the classification of the web attacks.

The role of the expert system is to eliminate the false alarms by consulting its knowledge data base, a task which is absent in rule-based IDS systems. Web attacks can be either rejected by the server or can be successful due to security weaknesses. If hacker intrusion occurs action must be taken by the security analyst as the prototype system does not deal with resolving the damage caused by an attack. It is solely a surveillance device.

Phase 4: Intelligent Visualization

This phase covers the last step of the Visual Analytics mantra. A user friendly visual tool had to present the security analyst with ongoing normal and malicious traffic. The output of the expert system should be transformed to a 3D graph for visual interpretation. Possible malicious traffic should be easily spotted together with normal traffic. To distinguish the different attack classes an eye-catching feature is necessary to help the analyst to recognise the attack and relate it to other suspect traffic. To accomplish this we used colouring for the different classes of attacks, choosing hot colours for the most dangerous attacks such as command or code injections.

Phase 5: Performance evaluation

Finally a method to measure the performance of the prototype system needed to be developed. For this purpose a statistics module had to be designed in order to analyse the classifier's behaviour. In this module statistics of following kind of traffic were kept:

1. Attacks present and rightly detected
2. Attacks absent but detected or misclassified attacks (False alarms)
3. Attacks present but not detected.
4. Normal traffic.

The rest of this chapter is organized as follows: section 2 presents related work, section 3 presents the modules of the visualization prototype ID system in details and section 4 describes the system's performance evaluation. Finally, concluding remarks appear in section 5.

7.2 Related Work

Interesting works on the detection of web-based attacks have been published in the last few years. Statistical methods have been used in [8] such as the multi-model approach for the detection of web-based attacks. A Bayesian parameter estimation technique for web session anomaly detection is described in [9] and DFA (Deterministic Finite Automata) induction has been applied in [10] to detect malicious web requests in combination with rules for reducing variability among requests and heuristics for filtering and grouping anomalies.

Recent works on application-level web security cover SQL and PHP code injections and XSS attacks. The authors in [11] combine a static analysis and runtime monitoring to detect and stop illegal SQL queries. In [12] a sound, automated static analysis algorithm is developed for the detection of injection vulnerabilities, modelling string values as context free grammars and string operations as language transducers. In [13] Noxes, a client-side solution is presented, which acts as a web proxy and uses both manual and automatically generated rules to mitigate possible cross-site scripting attempts. Additionally, in [14] Secubat, a web vulnerability scanner is described, which is a generic and modular web vulnerability scanner that, similar to a port scanner, automatically analyzes web sites with the aim of finding exploitable SQL injection and XSS vulnerabilities. Visual analytics have recently been applied in network monitoring [15], detection and analysis of protocol BGP anomalies [16] and Intrusion Detection [17].

Artificial Intelligence used for web intrusion detection is limited to Bayesian classifiers. In [18] an IDS system is presented based on a Bayesian classifier in the same vein as the now popular spam filtering software. A 2D tool named Bayesvis was implemented to apply the principle of interactivity and visualization to Bayesian intrusion detection. The tool reads messages as text strings and splits them up into the substrings that make the tokens. The major limitations of this system are the following: a) the training phase of the classifier is time-consuming as sufficient statistics for every type of web attack are needed for the efficient work of a Bayesian classifier. The training is also a laborious task as the operator has to perform manually the correction of false alarms. He/she starts by marking a few of the benign accesses and then he re-scores, re-sorts and repeats the process according to a predefined strategy, until the false positive rate arrives at an acceptable level, b) attacks against the web applications are not detected, such as backdoor intrusions and code injection attempts by high level applications such as SQL, Perl, Php, HTML and Java c) new attacks cannot be detected due to the absence of previous statistics d) only web logs, not real time web traffic, are processed.

Our work focused on creating an ongoing surveillance tool offering the security analyst a novel visual tool for monitoring and diagnostic needs. We would like to offer an online tool which is capable of dealing with real network traffic in addition to processing stored web logs. We used an unsupervised artificial neural network for grouping similar attacks into classes and an Evolutionary Artificial Neural Network for the web attack classification. In addition, we have expanded the signature method for ID to detect backdoor intrusions and code execution attempts by high level applications such as SQL, Perl, Php, HTML and Java. Attacks are classified automatically by the expert system, false alarms are very limited, new attacks not seen before are detected as well and simultaneous multiple attacks from different networks can be easily spotted on the screen from the IP source address labels and the colouring of the different attack classes. Additionally, the security analyst can examine in real time the malicious code of Perl, SQL or other high level language injections, Cross Site Scripting information and the code on new backdoor attempts such as worms and viruses.

Finally, we must emphasize that the whole system is developed in Linux and all system modules are written in standard C language, offering speed and portability to any operating system and platform, even on small portable computers.

7.3 Visualization Prototype System

Figure. 7.1 shows the architecture of the visualization prototype system.

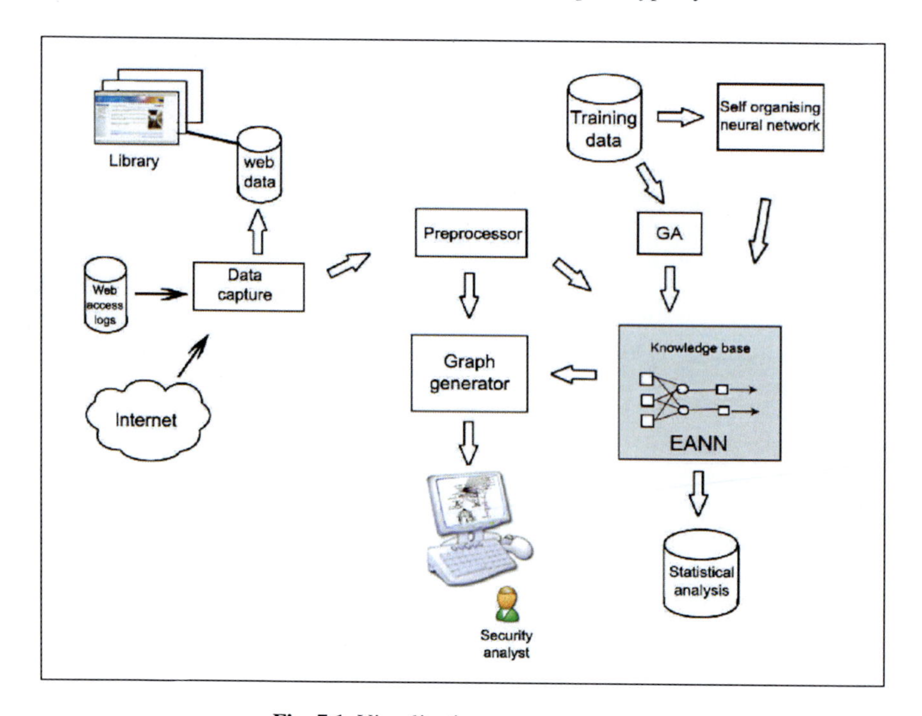

Fig. 7.1. Visualization prototype system

The visualization prototype system consists of the following modules:

- Data capture module,
- Pre-processor module,
- Knowledge base module,
- Graph generator module,
- Statistical analysis module.

The data capture module selects data either on-line from the Internet traffic or offline from the web server logs. The pre-processor module examines the web requests to detect malicious traffic and its output is then forwarded to the knowledge base module to predict the type of unauthorized traffic. Then, both normal and malicious traffic are processed by the graph generator module for visualization. Additionally, all traffic is processed for statistical analysis. Each module is described in detail below:

7.3.1 Data Capture Module

The two most popular web servers are Microsoft Internet Information Services (IIS) and the open source Apache web server. The IIS web server of the Library of the Technological Educational Institution (TEI) of Athens was used in order to study the various types of attacks and to create the knowledge data base of the system. Real data was captured with the tcpdump utility from June to end of November 2005. Using only real data we could not have a complete set of various attacks, so we have completed the tests with web logs data of the last three years. Web logs covered all versions of the Microsoft IIS server, e.g. V4 (Win NT 4.0), V5 (Win 2000), V6 and API (Win 2003). The size of real data was 95.298 web requests and the size of tested logs was 527.373, 620.033 and 23.577 events for the last three years respectively.

The logs of different IIS versions contain the same attributes but their syntax differs slightly from version to version. For instance, the web logs of V6 have the following structure:

#Software: Microsoft Internet Information Services 6.0
#Version: 1.0
#Date: 2005-06-25 13:37:21
#Fields: date time s-ip cs-method cs-uri-stem cs-uri-query s-port cs-username c-ip cs(User-Agent) sc-status sc-substatus sc-win32-status

Example:

2005-06-25 13:37:21 195.130.99.3 GET /cacti/image.php - 80 - 82.232.3.137 - 404 0 64

A web request captured online with the tcpdump utility has the following form:

IP 195.251.243.224.1412 > 195.130.99.96.80: tcp 268
GET /HM_Loader.js HTTP/1.1
Accept: */*
Referer: http://www.library.teiath.gr/
Accept-Language: el
Accept-Encoding: gzip, deflate
User-Agent: Mozilla/4.0 (compatible; MSIE 6.0; Windows 98; .NET CLR 1.1.4322)

Host: www.library.teiath.gr
Connection: Keep-Alive

From the above data the attributes to be processed by the system are the web request source IP (c-ip), the request command e.g GET (cs-method) and the request payload (cs-uri-stem).

7.3.2 Pre-processor Module

A total of 30 fingerprints was used in the model to group all the different types of known web attacks [19]. A detailed description of web attack fingerprints is given in [20]. Regular expressions were primarily used in the pre-processor module for the detection of fingerprints in the web requests or web logs. The presence of a fingerprint in a web request or log does not mean that it is a malicious attempt. Some fingerprints are used in web requests to launch a script on the server side and therefore these fingerprints are present in a normal request. The combination with other fingerprints allows the knowledge base module to identify a malicious attempt.

The pre-processor analyses the web request and creates a feature vector of dimension 30. Fingerprints are detected checking their decimal or hexadecimal representation. The presence of a specific fingerprint in the web request is indicated in the feature vector as 1 (true) and its absence as 0 (false or unknown). An attack may have more that one 1s fired in its vector representation and an attack belonging to a specific attack class has at least one binary representation. The output of the pre-processor module is two files, one with the feature vector and one with the web request data.

For instance the pre-processor for the following malicious web request:

00:25:37 213.23.17.133 - HEAD /Rpc/..%5c..%5c..%5cwinnt/system32/cmd.exe /c+dir+c:\ 404 143 99 0 HTTP/1.0 - - - produces the following two outputs:

1 1 0 0 0 0 0 0 0 0 0 0 1 0 0 0 0 0 0 0 0 0 0 0 0 0 0 0 0 0 (feature vector) and 213.23.17.133 HEAD /Rpc/..%5c..%5c..%5cwinnt/system32/cmd.exe /c+dir+c:\ (payload).

The feature vector will be the input to the expert system and the request data will be forwarded to the graph generator module. The extracted data from a web request are the most significant for the online analysis such as the source IP address, the request option (GET, HEAD etc.) and the request payload.

7.3.3 Knowledge Base Module

7.3.3.1 Classes of Web Attacks

Modern web servers offer optional features which improve convenience and functionality at the cost of increased security tasks. These optional features are taken into consideration in our design in addition to traditional types of web server attacks (Unicode, directory traversal, buffer overflow, Server-Side Includes-SSI, Cross Site Scripting-XSS, mail and CGI attacks). Attacks against web applications such as code injections or insertion attempts are detected in the following programming languages HTML, Javascript, SQL, Perl, Access and Php. In addition IIS indexing vulnerabilities, IIS highlight, illegal postfixes, IIS file insertion (.stm), IIS proxy attempts and IIS data access vulnerabilities (msadc) are detected as well. All .asa, .asp and Java requests are tested for URI (Uniform Resource Identifier) legal syntax according to standards, meaning that a corresponding query not in the form <?key=value> is illegal.

Trojan/backdoor upload requests are detected as well. These backdoors are left by worms such as Code Red, Sadmin/IIS and Nimda. Backdoor attempts for apache and IIS servers are detected when web requests ask for the corresponding password files (.sam and .htpasswd). Finally, command execution attempts are detected for both Windows (.exe, .bat, .sys, .com., .ini, .sh, .dll and other) and Unix (cat, tftp, wget, ls and other) environments.

To classify the above web attack types a self-organising neural network system has been used. The system was based on the famous Grossberg and Carperter's Adaptive Resonance Theory (ART1) [21]. ART1 algorithm is an unsupervised learning algorithm with biological motivations. Clustering algorithms are motivated by biology in that they offer the ability to learn through classification. Based on the Grossberg's *stability-plasticity dilemma* we cluster new concepts with older analogous ones and when we encounter new knowledge we create new clusters without destroying what has already been learned.

The ART1 neural network created 15 clusters or classes. These 15 classes were finally grouped manually into 9 as there was more that one class for command execution (Windows, Unix) and IIS type of attacks. It is interesting to notice that ART1 did not create a separate class for directory traversal and Unicode attacks because almost all of the web requests containing Unicode or traversal fingerprints (..\ or ../) always included another type of attack (e.g buffer overflow, command execution attempt, code insertions or other). So, directory traversal and Unicode attempts are not classified as separate attack classes. For historical reasons we included Unicode attempts into the Miscellaneous class.

The 9 final web attack classes used are the following:

1. *Commands (CMD)*: Unix or Windows commands for code execution attempts.
2. *Insertions (INS)*: Application code injections (SQL, Perl, HTML, Javascript, Data Access).
3. *Trojan Backdoor Attempts (TBA)*: Attacks triggered by virus and worms (Code Red II, Sadmin, Luppi etc.).
4. *Mail (MAI)*: Mail attacks through port 80 (formail, sendmail etc.).
5. *Buffer overflows (BOV)*: Attacks corrupting the execution stack of a web application.
6. *Common Gateway Interface (CGI)*: Exploitation of vulnerable CGI programs.
7. *Internet Information Server(IIS)*: Attacks due to vulnerabilities of IIS.
8. *Cross Site Scripting (XSS)* or *Server Side Includes (SSI)* attacks.
9. *Miscellaneous (MISC)*: Coldfusion, Unicode, and malicious web request options such as PROPFIND, CONNECT, OPTIONS, SEARCH, DEBUG, PUT and TRACE.

7.3.3.2 Training Data Quality

To measure the information which exists between the input data and the output data we had to calculate the *mutual information* between the two data sets. We want the network to take the input and remove all uncertainty about what the corresponding output should be. The amount of the original uncertainty we can remove depends on the mutual information present in the data. With an ideal training set, once we know the input value, there should be no doubt as to the correct output value: it should be the one value with a conditional probability, given the current input, of one; all other output values should have a probability of zero. As we cannot have an ideal training set, we need a measure of the average spread of conditional probabilities over the whole training set.

Let H denote the entropy of a set of events, X and Y the data sets of input and output respectively, H(X|Y) the conditional entropy of inputs given the outputs and I(X;Y) the mutual information between the input and the output data of the training set. We measured the H(X), H(Y) and I(X;Y) using a program to calculate the following equations:

$$H = \sum_{i=1}^{n} P_i * \log \frac{1}{P_i}$$

$$H(X \mid Y) = \sum_{i=1}^{n} \sum_{j=1}^{m} P(x_i, y_j) * \log \frac{P(y_j)}{P(x_i, y_j)}$$

$$I(X;Y) = H(X) - H(X \mid Y)$$

$$I(X;Y) = \sum_{i=1}^{n} \sum_{j=1}^{m} P(x_i, y_j) * \log \frac{P(x_i, y_j)}{P(x_i) * P(y_j)},$$

where n is the number of possible distinct input events, m the number of possible distinct output events and P_i is the probability of event i occurring out of the possible n events. Table 7.1 shows the results with the used training set.

As we can see: H(inputs) ≈ log(n) and H(outputs) ≈ log(m), so the used training set is a well balanced training set. The ratio I(input; output):H(output) ranges from 0 to 1 and would be high if a data set is learnable. This ratio for our data set is equal to 0.805, which means that the data set used is learnable. However, it could be improved in the future.

Table 7.1. Data sets entropy and mutual information results

n	log(n)	m	Log(m)	H(X)	H(Y)	H(X|Y)	H(Y|H)	I(X;Y)
49	3.891	9	2.197	3.512	2.160	1.777	0.420	1.740

7.3.3.3 Evolutionary Artificial Neural Network

If the pre-processor detects even one fingerprint its output is forwarded to an expert system for classification. In the first version of the prototype [22] we used an

Artificial Neural Network (ANN) for classification. ANN's represent a class of very powerful, general-purpose tools that have been successfully applied to prediction, classification and clustering problems. The ANN used was a multilayer network with one hidden layer, using the *generalized delta rule with the backpropagation (BP) algorithm* for learning and the sigmoid function as activation function [23]. The input neurons were 30 (+1 the bias), the hidden neurons 10 (+1 the bias) and the output neurons 9, representing the 9 web attack classes.

In the final version of the prototype a hybrid expert system is used for the web attacks classification. We used an Evolutionary Artificial Neural Network (EANN), which is neural network combined with Genetic Algorithms (GA) for weight optimization. GA's are algorithms for optimization and learning, based loosely on several features of biological evolution. GA's do not face the drawbacks of the backpropagation (BP) algorithm, such as the scaling problem and the limitation of the fitness (error) function to be differentiable or even continuous. If the problem complexity increases, due to increased dimensionality and/or greater complexity of data, the performance of BP falls off rapidly. GA's do not have the same problem with scaling as backpropagation. One reason for this is that they generally improve the current best candidate monotonically, by keeping the current best individual as part of their population while they search for better candidates. Secondly, they are not bothered by local minima.

Let us consider the three-layer neural network of the prototype system. The parameters *n, m* and *l,* denoting the number of neurons of the three layers, are respectively 30, 10 and 9 for the prototype system.

To find an optimal set of weights for the multilayer feedforward neural network we first need to represent the problem domain as a chromosome. Initial weights are chosen randomly within some small interval [-0.5, 0.5]. The set of weights can be presented by a square matrix (Figure 7.2) in which a real number corresponds to the weighted link from one neuron to another and zero means that there is no connection between two given neurons. Since a chromosome is a collection of genes, a set of weights can be represented by an n-gene chromosome, where each gene corresponds to a single weighted link in the network. Thus, if we string the rows of the matrix together, ignoring zeros, we obtain a chromosome.

Each row of the matrix represents a group of all the incoming weighted links to a single neuron. This group can be thought of as a functional building block of the network [24] and therefore should be allowed to stay together passing genetic material from one generation to the next. To achieve this we associated each gene of the chromosome not with a single weight but with a group of weights, a row of the above matrix.

The number of neurons are the same as in the original neural network. So, as in total there are 409 weighted links (31*10 + 11*9) between neurons, the chromosome has a dimension of 409 and a population member has been represented as:

$$\mathbf{M} = \langle \mathbf{w}_{0,0}, \mathbf{w}_{1,0} \ldots \mathbf{w}_{30,0}, \mathbf{w}_{0,1}, \mathbf{w}_{1,1} \ldots \mathbf{w}_{30,1}, \ldots\ldots, \mathbf{w}_{0,9}, \mathbf{w}_{1,9} \ldots \mathbf{w}_{30,9} \mid \mathbf{w}_{0,0}, \mathbf{w}_{1,0} \ldots \mathbf{w}_{10,0}, \mathbf{w}_{0,1}, \mathbf{w}_{1,1} \ldots \mathbf{w}_{10,1}, \ldots\ldots, \mathbf{w}_{0,8}, \mathbf{w}_{1,8} \ldots \mathbf{w}_{10,8} \rangle,$$

where, the first part is the transposed matrix $W_{ih}[31,10]$ of weights between the input and the hidden layer (matrix A in Figure 7.2, we string the rows together) and the second part concatenated is the transposed matrix $W_{ho}[11,9]$ of weights between the

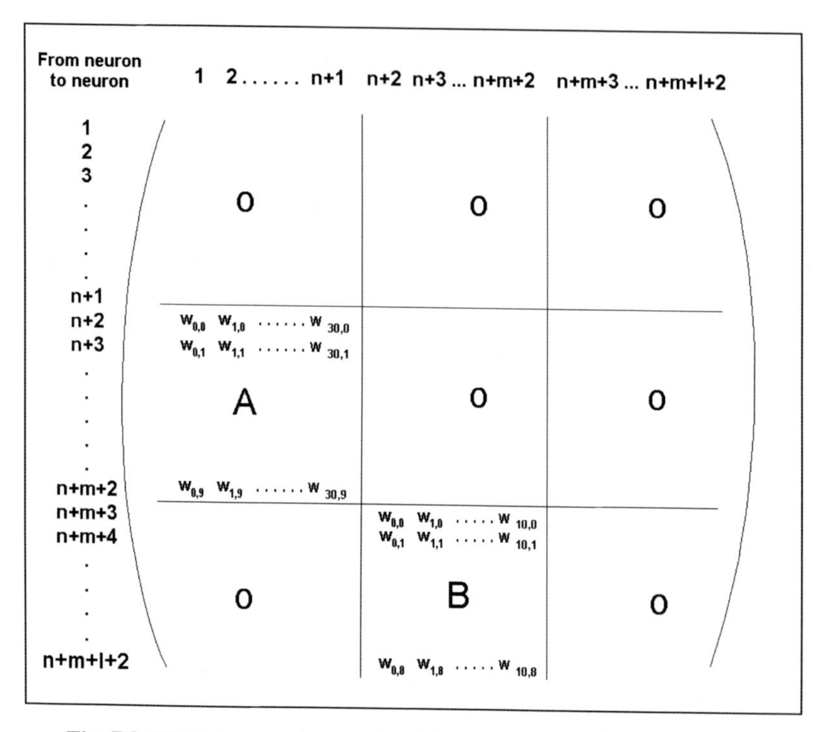

Fig. 7.2. Weight connection matrix of the three layer (BP) neural network

hidden layer and the output (matrix B in Figure 7.2). Each member of the population was coded with the structure of the chromosome and a double real number for the fitness number.

The fitness function for evaluating the chromosome's performance was the sum of squared errors (SSE), used in the training phase of the BP algorithm. The smaller the sum, the fitter the chromosome.

We used crossover and mutation as genetic operators. A crossover operator takes two parent chromosomes and creates a single child with genetic material from both parents. Each gene in the child's chromosome is represented by the corresponding gene of the randomly selected parent. A mutation operator randomly selects a gene in a chromosome and adds a small random value between -0.5 and 0.5 to each weight in this gene. The crossover and mutation probabilities were 0.8 and 0.05 respectively. Firstly a mutation probability of 0.02 was used, but finally it raised to 0.05, as it accelerated the evolution of the GA.

The used algorithm of the EANN system can be described in a pseudo-code as following:

1. Randomly generate an initial population of chromosomes (population size 30) with weights in the range of [-0.5, 0.5].
2. Train the network for 1000 epochs using the BP algorithm. Calculate the fitness function for all individuals.
3. Select a pair of chromosomes for mating with a probability proportional to their fitness (roulette-wheel selection).

4. Create a pair of offspring chromosomes by applying the genetic operators crossover (multi-point crossover) and mutation.
5. Place the created offspring chromosomes in the new population.
6. Repeat step 4, until the size of the new population becomes equal to the size of the initial population, and then replace the parent chromosome population with the new (offspring) population.
7. Go to step 2 and repeat the process until the algorithm converges or a specified number of generations has been reached (we used a maximum of 1000 generations).
8. Use the weights of the best member (ideal) of the last generation for the feedforward only operation of the ANN (classification).

7.3.3.4 EANN Performance Versus ANN

For each generation the minimum (minFit) , the average (avgFit) and the maximum fitness (maxFit) of the population were calculated, in the belief that the algorithm should converge if the minimum fitness was less than an epsilon, equal to 10^{-12} and the ratio minFit/avgFit was greater that 0.95. In this way, setting such a severe criterion all members of the final generation would become "ideal" and fit to be used for classification in the feedforward neural network, not just the best member of the population.

The algorithm did converged after 305 generations (Figure 7.3) giving a minimum fitness of 6.61e-12 and 30 ideal members, a set of 30 best optimized weights for the operation of the ANN. Figure 7.4 shows the performance of the EANN hybrid expert system versus a simple ANN.

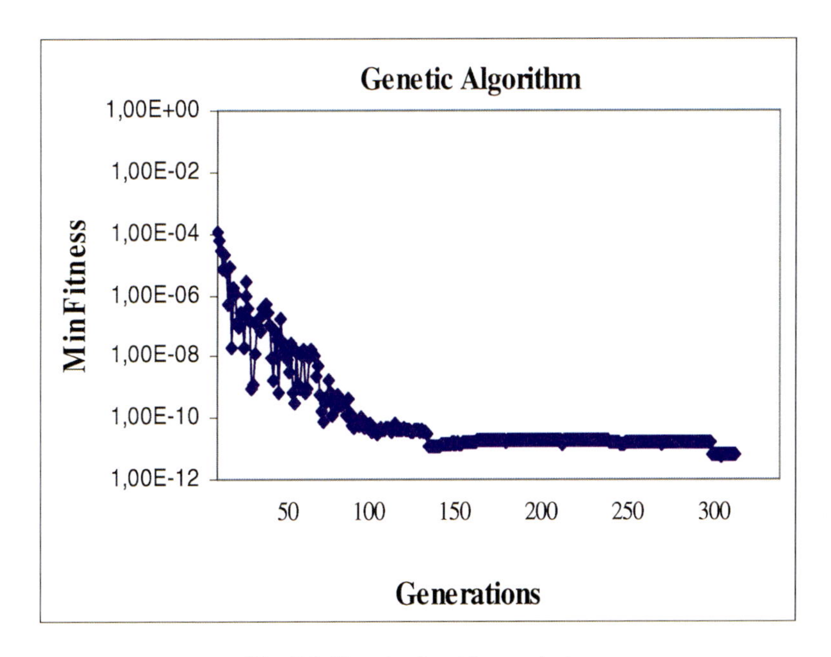

Fig. 7.3. Genetic algorithm evolution

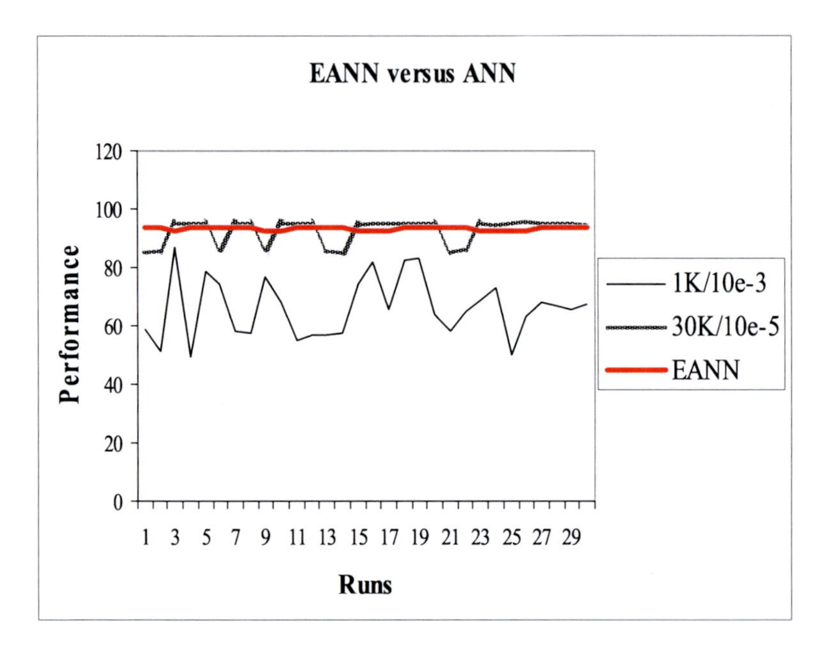

Fig. 7.4. EANN performance versus ANN (test data)

In Figure 7.4 the straight line indicates the stable performance (95.70%) of the EANN using a set of test data for the performance test (not the training set). Initial training was done with only 1000 epochs and a SSE limit of 10^{-3}. The other two lines show the performance of a simple ANN using the BP algorithm. We can distinguish the stochastic behavior of the ANN's performance. Using 1000 epochs and a SSE limit of 10^{-3} the ANN system performance rated between 50-87%, giving an average of 66.15% for 30 runs. Using 30,000 epochs and a SSE limit of 10^{-5} the ANN system performance rated between 85-95% giving an average of 92.52% for 30 runs. In the first version of the prototype system the latter combination was used, which had the drawback of a slower training cycle.

Using a test data set and the hybrid expert system with the GA approach for the weight optimization a stable neural network performance of 93.51% was achieved for all the 30 runs (red straight line in Figure 7.4).

7.3.4 Graph Generator Module

The predicted attack by the EANN is then used to create a coloured directed graph in *dot* form of the well known GraphViz [25] package, using the corresponding *DOT* language. This language describes four kinds of objects: graphs, nodes, edges and labels and has a large number of attributes that affect the graph drawing.

The payload of a web request is cut into nodes and the directed edges are the links between these nodes from left to right. Therefore, a web request from an IP source 217.229.196.17 with payload GET /hact/graphics/springer.gif, has as nodes the words

"217.229.196.17", "GET", "hact", "graphics", "springer.gif" and as "directed edges" the links between these nodes from left to right:

217.229.196.17 → GET → hact → graphics → springer.gif.

When each web request with its IP source address and the requested data is visualized in a 3D graph the security analyst can navigate into the graph for a quick interpretation and evaluation in case of a malicious attempt. Timestamps were not added to the graph as graphs are displayed in real time and the objective here is to keep the display as simple as possible.

There are two graphs generated with the GraphViz package. One graph contains real time traffic, e.g. both normal and possible malicious traffic and the other does not contain normal but only possible malicious traffic. Normal traffic is visualized in black and malicious traffic in 9 different colours, one for each attack class, such as red (Commands), brown (Insertions), magenta (Backdoor attempts), green (Mail), cyan (Buffer overflows), gold (CGI), blue (IIS), yellow (XSS) and coral (Miscellaneous). This visual separation was necessary because normal traffic overloads the display and the security analyst cannot interpret quickly the malicious attempts. When visualizing both normal and malicious traffic the security analyst spends more time navigating through the graph trying to eliminate normal traffic by zooming into the coloured part of the display, than he would if he had only a coloured graph to contend with.

The malicious traffic colour display in both graphs is the result of the visual analytics process whereby web traffic is first analysed (i.e., the most important data is chosen, Intrusion Detection analysis is done using Artificial Intelligence) and then displayed. By employing intelligent means in the analysis process the visual representation of the traffic allows the security analyst to gain insight into the intrusion problem quickly and will be of invaluable help in the decision making process as he will be able to rapidly extract knowledge from the graph.

These two *dot* coloured graphs are then visualized with Tulip [26], a 3D graph visualization tool, supporting various graph algorithms and extensive features for interactive viewing and graph manipulation.

Figures 7.5, 7.7, 7.9 show normal and malicious web traffic and Figures 7.6, 7.8, 7.10 only the malicious traffic for the same events.

In Figure 7.6 the cyan graph indicates a buffer overflow (the character "d" repeated more than 200 times) from IP 195.130.99.100, the green graph a formail attempt from IP 195.130.99.218, the blue graph an IIS attempt, the brown an insertion attempt, the red graph a command execution attempt and the magenta graph a Trojan backdoor attempt.

In Figure 7.8 the brown graph shows a backdoor attempt (perl injection) with the recent Linux/Lupper worm aka luppi worm. The latter is a new attack which appeared in November 2005 and was detected by the system which was not trained for this kind of code insertion.

In Figure 7.9 the brown graphs in the right indicate simultaneous Perl injection attempts from IP 195.102.4.156 and 211.189.119.85, the red graphs indicate multiple command execution attempts from IP 200.24.5.98 and other sources and the magenta graphs indicate multiple backdoor attempts (Code Red II) from IP 217.229.196.17.

Finally, in Figure 7.10 we can spot additional command execution attempts from IP 213.23.17.133 and buffer overflows attacks from IP 195.77.248.102 (cyan graph). The Perl injection code can be easily read on the right bottom of the graph.

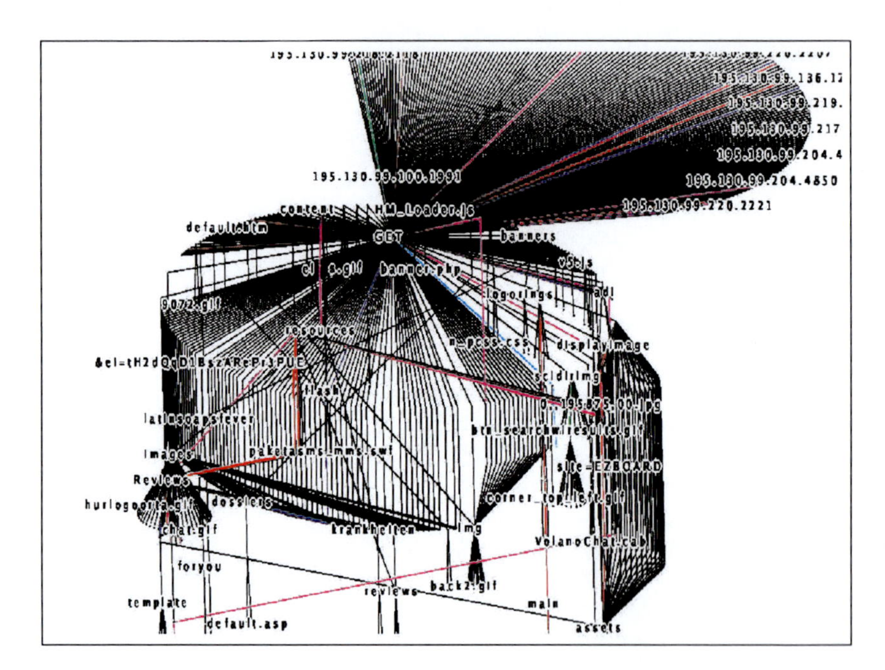

Fig. 7.5. Normal and malicious traffic (online data 14/6/2005)

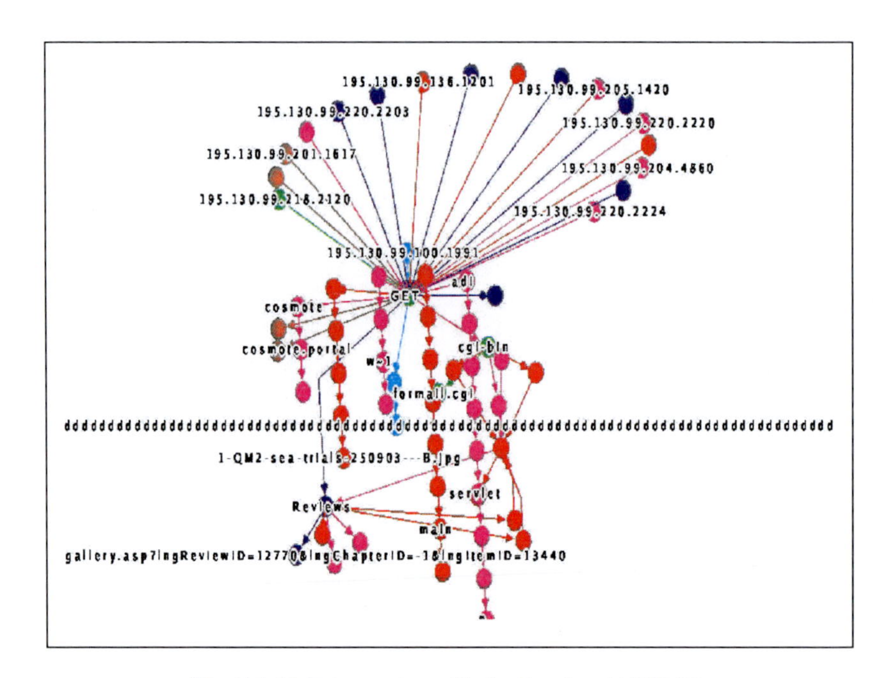

Fig. 7.6. Malicious only traffic (online data 14/6/2005)

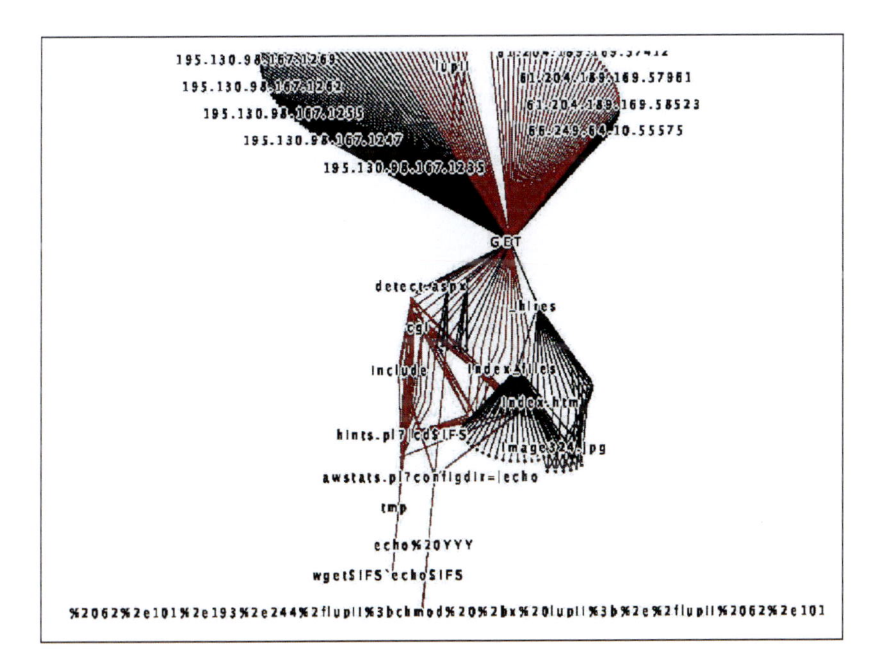

Fig. 7.7. Normal and malicious traffic (online data 9/11/2005)

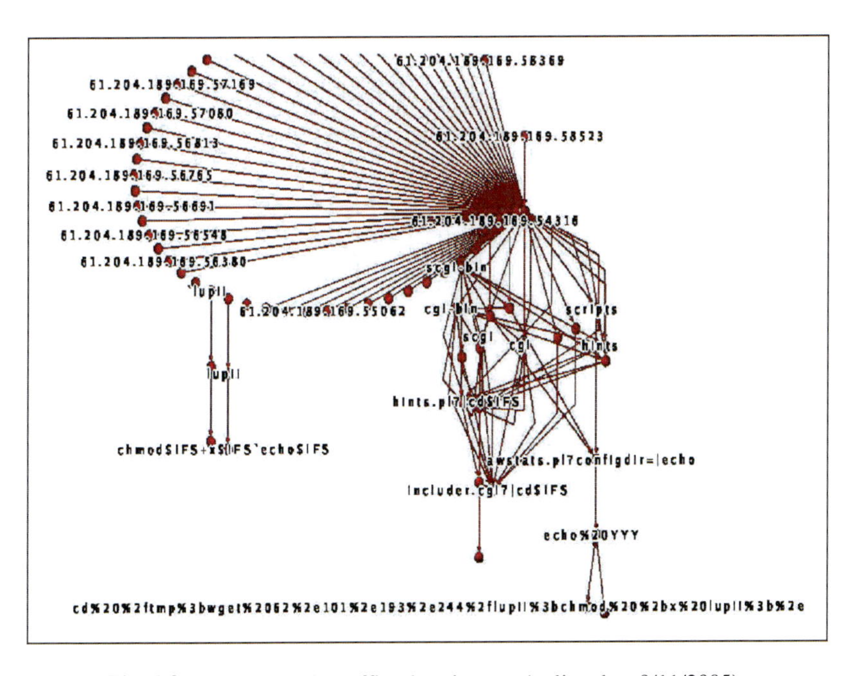

Fig. 7.8. Malicious only traffic - luppi worm (online data 9/11/2005)

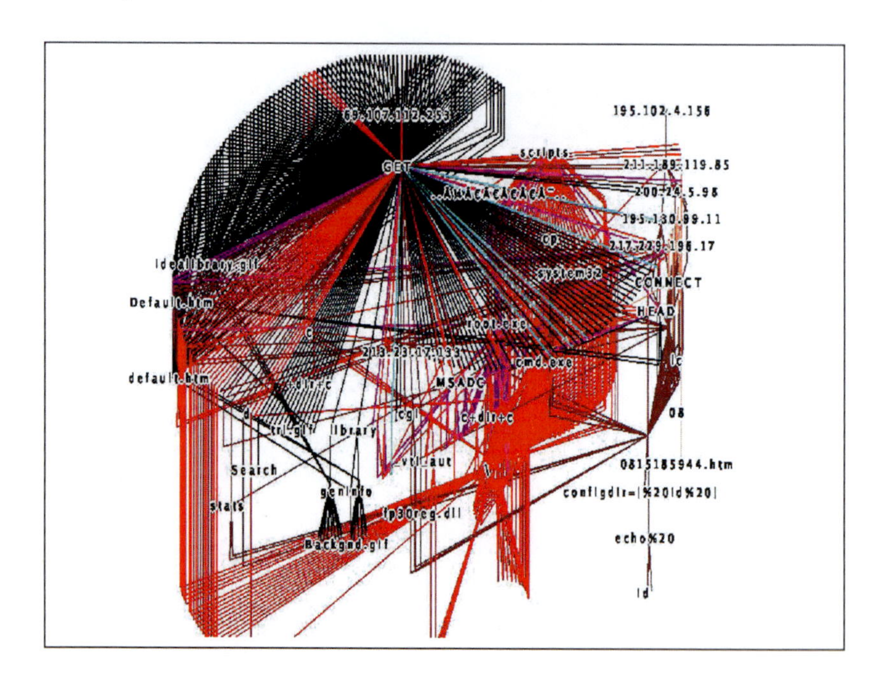

Fig. 7.9. Normal and malicious traffic (web logs 2006)

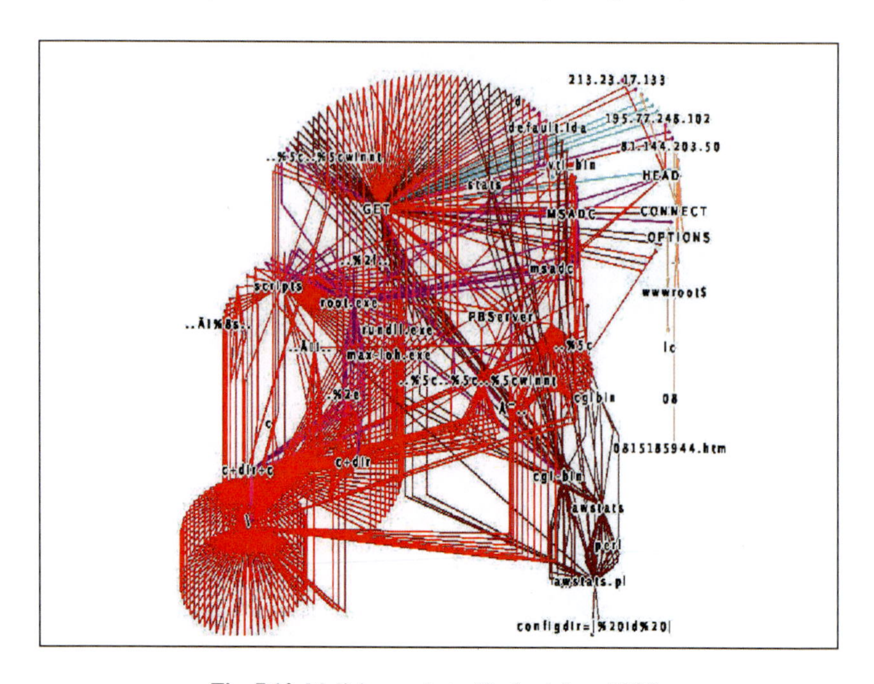

Fig. 7.10. Malicious only traffic (web logs 2006)

7.3.5 Statistical Analysis Module

The system's performance was tested using real data, captured with the *tcpdump* utility in June and November 2005 and web logs of 2005 and 2006. For each test a 2x2 table is calculated containing, on the first row the Hits (attacks present or True Positives) and the False Alarms (or False Positives) and on the second row the Misses (attacks present but not detected or False Negatives) and the Correct Rejections (normal traffic or True Negatives). Results are presented in Table 7.2 in this form. All tests have been run for various values of a detection threshold to show how changing the detection threshold affects detections versus false alarms. If the threshold is set too high then the system will miss too many detections and conversely, if the threshold is set too low there will be too many false alarms. For the tests we have used threshold values rating from 0.3 to 1.0 with a step of 0.1. The best results using the BNN were obtained with a threshold value of 0.8 giving maximum detections of 95% and a minimum of false alarms. Using the EANN we obtained almost the same results for a threshold rating between 0.3 and 0.9, due to the stable performance (93.50%) of the hybrid expert system. Table 7.2 summarizes the results with various testing data sets.

In addition, in the statistical analysis module of the system a confusion matrix is calculated to display the classification results of a network. The confusion matrix is defined by labelling the desired classification in rows and the predicted classifications in columns. For each exemplar, a 1 is added to the cell entry defined by (desired

Table 7.2. Performance evaluation tests (EANN)

Test Data Positives Negatives	Logs 2005 149545 events	Logs 2005 149456 events	Logs 2006 149450 events	Logs 2006 149503 events	Logs 2006 149749 events	online data (Oct. 05) 49372 events	online data (Nov. 05) 22022 Events
TP FP	18485 255	9136 12	7575 2	10176 0	3631 0	9 0	34 22
FN TN	25 130780	582 139726	56 141817	62 139265	63 146055	0 49363	13 21953

Table 7.3. Confusion matrix for test1 (EANN with threshold 0.7)

	CMD	INS	TBA	MAI	BOV	CGI	IIS	XSS	MIS	NRM
CMD	**17469**	**241**	0	0	0	0	0	**9**	0	0
INS	0	**5**	0	0	0	0	0	0	0	0
TBA	0	0	**312**	0	0	0	0	0	0	0
MAI	0	0	0	**3**	0	0	0	0	0	0
BOV	0	0	0	0	**421**	0	0	0	0	0
CGI	0	0	0	0	0	**7**	0	0	0	0
IIS	0	0	0	0	0	0	**95**	0	0	0
XSS	0	**5**	0	0	0	0	0	**0**	0	0
MIS	0	0	0	0	0	0	0	0	**173**	0
NRM	0	0	0	0	0	0	0	0	0	**130780**

Hits: 18485 False Alarms: 255
Missed: 25 Normal traffic: 130780 Total events: 149545

classification, predicted classification). Since we want the predicted classification to be the same as the desired classification, the ideal situation is to have all the exemplars end on the diagonal cells of the matrix. Table 7.3 shows such a confusion matrix for test1 (web logs 2005).

7.4 Prototype System Performance

7.4.1 Introduction

There are two main divisions of classification: *supervised classification* (or *discrimination*) and *unsupervised classification* (or *clustering*). In supervised classification we have a set of data samples, each consisting of measurements on a set of variables, with associated labels, the class types. These are used as exemplars in the classifier design. In unsupervised classification, the data are not labelled and we seek to find groups in the data and the features that distinguish one group from another.

There are two approaches to measure the supervised classification. The first assumes a knowledge of the underlying class-conditional probability density functions e.g. the probability density function of the feature vectors for a given class. The second approach develops rules that use the data to estimate the decision boundaries directly, without explicit calculation of the probability density function.

These two approaches are the Bayes' decision rule and the Neyman-Pearson decision rule. The first is a theoretical approach to performance measurement and it cannot be used for our prototype system, as it presumes that the probabilities of each class occurring (*a-priori* probabilities), are known.

In the context of the Neyman-Pearson approach we have calculated the Receiver Operating Characteristic (ROC) for the prototype system, as a means of characterizing its performance. ROC provides a good means of visualizing the prototype's performance in order to select a suitable decision threshold. The ROC curve is a plot of the true positive rate on the vertical axis against the false positive rate on the horizontal axis. In the terminology of signal detection theory, it is a plot of the probability of detection against the probability of false alarm, as the detection threshold is varied. The performance of the prototype IDS system has been evaluated in [27].

7.4.2 Classification

7.4.2.1 Neyman-Pearson Decision Rule

An alternative to the Bayes' decision rules for a two class problem is the Neymann-Pearson test [28]. In a two-class problem there are two possible types of error that may be made in the decision process. We may classify a pattern of class ω_1 as belonging to class ω_2 or a pattern from class ω_2 as belonging to class ω_1. Let the probability of these two errors be ε_1 and ε_2 respectively, so that

$$\varepsilon_1 = \int_{\Omega_2} p(x \mid \omega_1) dx \qquad \text{error probability of Type I}$$

and

$$\varepsilon_2 = \int_{\Omega_1} p(x \mid \omega_2)dx \qquad \text{error probability of Type II}$$

If class ω_1 is termed the positive class and class ω_2 the negative class, then ε_1 is referred to as the *false negative rate*, the proportion of positive samples incorrectly assigned to the negative class and ε_2 is the *false positive rate*, the proportion of negative samples classed as positive.

If ω_1 denotes the signal probability and ω_2 denotes the "noise" (term used in signal theory) then ε_2 is the probability of false alarms (P_F) and ε_1 is the probability of missed detections (P_M). In many applications a threshold is set to give a fixed probability of false alarms.

The Neyman-Pearson decision rule is to minimize the error ε_1 subject to ε_2 being equal to a constant, a, say. Using different terminology, the Neyman-Pearson decision rule is to maximize the detection probability P_D (P_D=1-ε_1), while not allowing the false alarm probability (P_F) to exceed a certain value.

7.4.2.2 Sufficient Statistics and Monotonic Transformations

Consider the test

$H_0 : x \sim f_0(x)$
$H_1 : x \sim f_1(x)$, where $f_i(x)$ is a density.

The solution to the optimization problem is given by

$$L(x) = \frac{f_1(x)}{f_0(x)} \underset{H_0}{\overset{H_1}{\gtrless}} \gamma,$$

where $L(x)$ is the **likelihood ratio**, and γ is a threshold.

γ is such that $P_F = \int_{\forall x, L(x)>\gamma} f_0(x)dx = \alpha$. The detection probability is $P_D = \int_{\Omega_1} f_1(x)dx$.

The optimal decision rule is called the **Likelihood Ratio Test (LRT)**. The threshold can often be solved for as a function of α.

The densities $f_i(x)$ are non-negative, so as Ω_1 shrinks, both probabilities tend to zero. As Ω_1 expands, both tend to one. The ideal case, where $P_D = 1$ and $P_F = 0$, cannot occur unless the distributions do not overlap (i.e., $\int f_0(x)f_1(x)dx=0$). Therefore, in order to increase P_D, we must also increase P_F. This represents the fundamental tradeoff between hypothesis testing and detection theory.

For hypothesis testing involving multiple or vector-valued data, direct evaluation of the size (P_F) and power (P_D) of a Neyman-Pearson decision rule would require integration over multi-dimensional, and potentially complicated decisions regions. However, in many cases this can be avoided by simplifying the likelihood ratio test to a test of the form

$$t \underset{H_0}{\overset{H_1}{\gtrless}} \gamma$$

, where the test statistic $t = T(x)$ is a sufficient statistic for the data.

Such a simplified form is arrived at by modifying both sides of the likelihood ratio test with monotonically increasing transformations and by algebraic simplifications. Since the modifications do not change the decision rule, we may calculate P_F and P_D in terms of the sufficient statistics. Thus, the false-alarm probability may be written:

$$P_F = P_r\,[\text{declare } H_1] = \int_{\forall t, t>\gamma} f_0(t)dt, \qquad \text{where}$$

$f_0(t)$ denotes the density of t under H_o. Since t is typically of lower dimension than x, evaluation of P_F and P_D can be greatly simplified. The key is being able to reduce the likelihood ratio test involving a sufficient statistic for *which we know the distribution.*

7.4.2.3 Neyman-Pearson Lemma: General Case

Let Φ be a function of the data x with $\Phi(x) \in [0,1]$. Φ defines the decision rule "declare H_1 with probability $\Phi(x)$".

Consider the hypothesis testing problem:

$H_0 : x \sim f_0(x)$
$H_1 : x \sim f_1(x)$, where f_o and f_1 are both density functions.

Let $a \in [0,1)$ be the size constraint (false-alarm probability). The decision rule

$$\Phi(x) = \begin{cases} 1 & \text{if } L(x) > \gamma \\ \rho & \text{if } L(x) = \gamma \\ 0 & \text{if } L(x) < \gamma \end{cases}$$

is the most powerful test of size α [29], where γ and ρ are uniquely determined by requiring $P_F = \alpha$. If $\alpha = 0$, we take $\gamma = \infty$, $\rho = 0$.

When $P_r[L(x)] = \gamma] > 0$ for certain γ, we choose γ and ρ as follows:

$P_r[L(x) > \gamma] \le \alpha \le P_r[L(x) \ge \gamma]$ and
$\rho\, P_r[L(x) = \gamma] = \alpha - P_r[L(x) > \gamma]$.

The false alarm probability is: $P_F = P_r[L(x) > \gamma] + \rho\, P_r[L(x) = \gamma]$.

7.4.3 Detection, False and Miss Probabilities of the Prototype System

If the predicted vector of the classifier belongs to one of the C possible classes C_i, i = 1...C, then it is assumed that the predicted attack belongs to class C_i. So, a one

(1) is received at the position *i*, indicating the corresponding class C_i and all the other components of the vector are zero (0), i.e. if the predicted vector is (0,1,0...0) it means that the predicted attack belongs to class C_2. Supposing, that the classifier runs N times trying to classify the same event it does not produce the same vector every time, due to classification errors (or noise), e.g. it does not produce a one (1) N times at the same position, indicating the class C_i. Assuming the N runs are statistically independent, the values we receive are Bernoulli random variables: $x_n \sim$ Bernoulli (θ).

We are faced with the following hypothesis test:

$H_o: \theta = p$ (0 output)
$H_1: \theta = 1\text{-}p$ (1 output), where
p is the probability that a value is flipped ($0 \leftrightarrow 1$), $0 \le p < 0.5$ and p is known.

For a certain class C_i the received sequence will be decoded $\boldsymbol{x} = (x_1, x_2, \ldots x_N)^T$ by designing a Neyman-Pearson rule.

The join density of x_n is :

$$f(x_{1,\ldots}x_N) = \prod_{i=1}^{N} p^{X_i}(1-p)^{1-X_i} = p^{\sum X_i} * (1-p)^{N-\sum X_i} = p^k * (1-p)^{N-k},$$

$$\left(k = \sum_{i=1}^{N} x_i\right),$$

where k is the number of 1s received.

The conditional probability of $\boldsymbol{x}$ given k is independent of θ as:

$$f_{\theta|k}(x) = \frac{f_\theta(x,k)}{f_\theta(k)} = \frac{\theta^k(1-\theta)^{N-k}}{\binom{N}{k}\theta^k(1-\theta)^{N-k}} = \frac{1}{\binom{N}{k}},$$

so, k is a sufficient statistic for θ and the N values $x_1, x_2, \ldots x_N$ can be replaced by the low-dimensional quantity k without losing information about θ.

The likelihood ratio is:

$$L(x) = \frac{(1-p)^k p^{N-k}}{p^k(1-p)^{N-k}} = \left(\frac{(1-p)}{p}\right)^{2k-N}$$

The LRT is $$\left(\frac{1-p}{p}\right)^{2k-N} \underset{H_0}{\overset{H_1}{\gtrless}} t$$

By taking the logarithms of both sides, we have:

$$k \underset{H_0}{\overset{H_1}{\gtrless}} \frac{N}{2} + \frac{1}{2} \frac{\ln t}{\ln\left(\frac{1-p)}{p}\right)} = \gamma$$

The false alarm probability is

$$P_F = P_r[k > \gamma] + \rho P_r[k = \gamma] = \sum_{k=\gamma+1}^{N} \left(\binom{N}{k} p^k (1-p)^{N-k} \right) + \rho \binom{N}{\gamma} p^{\gamma} (1-p)^{N-\gamma} \quad (7.1)$$

where γ and ρ are chosen so that $P_F = \alpha,$ as described above.

The corresponding detection probability is

$$P_D = P_r[k > \gamma] + \rho P_r[k = \gamma] = \sum_{k=\gamma+1}^{N} \left(\binom{N}{k} (1-p)^k P^{N-k} \right) + \rho \binom{N}{\gamma} (1-p)^{\gamma} p^{N-\gamma} \quad (7.2)$$

7.4.4 ROC Curve of the Prototype System

Running the system N times ($N = 10$) to classify the same attack we measured p=0.3. When a zero is received (instead of a one) it means that the classifier either misclassifies the attack or misses to detect the specific attack.

From the previous equation of P_F (7.1) we calculate γ and ρ so that $P_F = a$ (Table 7.4).

Table 7.4. Threshold values γ versus different interval values of $P_F(\alpha)$

γ	$P_r[k=\gamma]$	$P_r[k>\gamma]<\alpha\leq P_r[k\geq\gamma]$
γ=8	0,013	$0.0001< \alpha \leq 0.013$
γ=7	0.008	$0.013 < \alpha \leq 0.021$
γ=6	0.035	$0.021 < \alpha \leq 0.056$
γ=5	0.084	$0.056 < \alpha \leq 0.140$
γ=4	0.196	$0.140 < \alpha \leq 0.336$
γ=3	0.265	$0.336 < \alpha \leq 0.601$
γ=2	0.230	$0.601 < \alpha \leq 0.831$
γ=1	0.120	$0.831 < \alpha \leq 0.951$
γ=0	0.028	$0.951 < \alpha \leq 0979$

For a given value of P_F and γ we calculate ρ as following:

Example: For $P_F = a = 0{,}1$ $\quad \rho = \frac{a - P_r[k>5]}{P_r[k=5]} = \frac{0.10 - 0.056}{0.084} = 0.523$

To find the detection probability P_D we first calculated the following table (Table 7.5):

Table 7.5. Threshold values γ for the computation of P_D

γ	$P_r[k=\gamma]$	$P_r[k>\gamma]$
$\gamma=3$	0.008	0.958
$\gamma=4$	0.035	0.923
$\gamma=5$	0.084	0.839
$\gamma=6$	0.196	0.643
$\gamma=7$	0.265	0.378
$\gamma=8$	0.230	0.148
$\gamma=9$	0.120	0.028

Then, from equation (7.2) we calculated P_D as following:

$$P_D = P_r[k>5] + \rho * P_r[k=5] = 0.839 + 0.523 * 0.084 = 0.882$$

Repeating the calculus for different values of $P_F = \alpha$ we filled up the following table (Table 7.6) for fault (P_F), detection (P_D) and miss ($1\text{-}P_D$) probabilities:

From the results of Table 7.6 we verified that in order to increase the P_D we must also accept an increase of P_F.

Table 7.6 can be interpreted as following:

If after 10 runs *(N=10)* of the classifier for the same event containing an attack, for example, more that 5 one's or successes ($k > 5$) are received, then with a false

Table 7.6. Fault, Detection and Miss probabilities of the prototype system

$P_F = \alpha$	γ	ρ	P_D	$1\text{-}P_D$
0.05	6	0.828	0.8053	0.1947
0.10	5	0.523	0.8829	0.1171
0.14	5	1.000	0.9230	0.0770
0.20	4	0.306	0.9337	0.0663
0.25	4	0.561	0.9426	0.0574
0.30	4	0.816	0.9515	0.0485
0.35	3	0.052	0.9584	0.4160
0.40	3	0.241	0.9599	0.0401
0.45	3	0.430	0.9614	0.0386
0.50	3	0.618	0.9629	0.0371
0.55	3	0.807	0.9644	0.0356
0.60	3	0.996	0.9659	0.0341
0.65	2	0.213	0.9662	0.0338
0.70	2	0.430	0.9666	0.0334
0.75	2	0.647	0.9666	0.0334
0.80	2	0.865	0.9668	0.0332
0.85	1	0.158	0.9670	0.0330
0.90	1	0.575	0.9672	0.0328
0.95	1	0.991	0.9674	0.0326

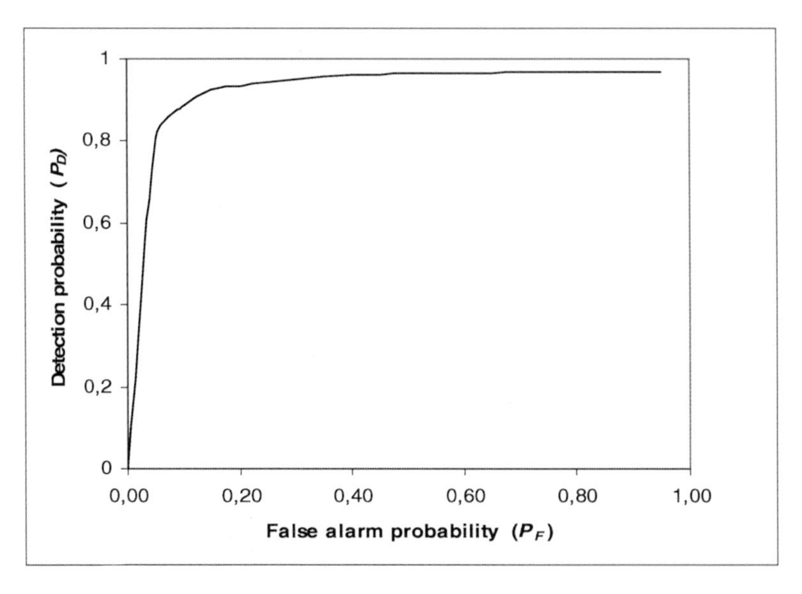

Fig. 7.11. Receiving Operating Characteristic (ROC) curve of the prototype system

probability of 14% there is a detection probability of 92.3% and a miss probability of 7.7% (line 3 of Table 7.6).

Finally, Figure 7.11 displays the ROC curve for the prototype system using the results of Table 7.6. From the ROC curve this can be verified visually as well, that with a P_F of around 15% a maximum detection (P_D) of 92% is achieved (upper left point of the curve). This is the best tradeoff between the false alarm rate and the detection rate of the developed prototype system.

7.5 Conclusion

It is technologically impossible for any device to understand application communications or analyse application behaviour through deep inspection of IP packets, either individually or reassembled into their original sequence. Network firewalls and Intrusion Detection Systems (IDS) are useful for validating the format of application header information to ensure standards compliance. In addition, network-level security devices may detect a small number of known, easily identifiable attacks by looking for pre-programmed patterns (i.e. attack signatures) in an HTTP stream.

Unfortunately, without any awareness of the HTML data payload or session context, devices that rely exclusively on the inspection of IP packets will fail to detect the vast majority of application-layer exploits. For example, IP packet inspection will not detect a hacker who has maliciously modified parameters in a URL (Universal Resource Locator) request.

Network data analysis is a very important but a time consuming task for any administrator. A significant amount of time is devoted to sifting through text-only log files and messages generated by networks tools in order to secure networks. Artificial intelligence and visualization offer a powerful means of analysis that can help the security analyst uncover hacker trends or strategies that are likely to be missed with other non-visual methods.

With our work we have contributed the following to artificial intelligence and network security:

- Use of an Evolutionary Neural Network as a knowledge base for rapid classification of web attacks. The stable performance of the EANN establishes it as a better classifier for web intrusion than a simple neural network.
- The application of automatic analysis methods before the interactive visual representation offers an intelligent visualization of web traffic that enables rapid perception and detection of unauthorized traffic.
- A surveillance aid for the security analyst, a strong tool for decision-making.
- A visualization prototype system ideal for educational purposes and in understanding web server and web application security.

This project has demonstrated that artificial intelligence considerably reduces the time required for data analysis and at the same time provides insights which might otherwise be missed during textual analysis. The web traffic surveillance could be expanded to other basic but popular internet services, such as email or DNS.

Combining traditional or novel analytical methods with visual presentation techniques can generate a very robust approach to network security. Artificial intelligence and visual analytics can be incorporated in ID systems to produce more powerful security systems capable of dealing with new attack challenges and noisy data. This is undoubtedly the future in the ID area.

References

[1] CVE, Common Vulnerabilities and Exposures, The Standard for Information Security Vulnerability Names (2008), http://www.cve.mitre.org

[2] Cobb, M.: Software security flaws begin and end with web application security (2008), http://searchsecurity.techtarget.com

[3] Snort software (2008), http://www.snort.org

[4] Komlodi, A., Goodall, J.R., Lutters, W.G.: An Information Visualization Framework for Intrusion Detection. In: CHI 2004 extended abstracts on Human factors in computing systems, Vienna, Austria, pp. 1743–1746. ACM press, New York (2004)

[5] Andreinko, G., Keim, D.A.: European Research Forum Panel Session: Envisioning Research Challenges in Visual Analytics. In: Proceedings of the 10th International Conference on Information Visualization (IV 2006), London, UK, pp. 5–7 (2006)

[6] Thomas, J., Cook, K.A.: A Visual Analytics Agenda. IEEE Transactions on Computer Graphics and Applications 26(1), 12–19 (2006)

[7] Keim, D.A., Mansmann, F., Schneidewind, J., Ziegler, H.: Challenges in Visual Data Analysis. In: Proceedings of the 10th International Conference on Information Visualization (IV 2006), London, UK, pp. 9–14 (2006)

[8] Kruegel, C., Vigna, G., Robertson, W.: A multi-model approach to the detection of web-based attacks. Computer Networks 48(5), 717–738 (2005)

[9] Cho, S., Cha, S.: SAD: web session anomaly detection based on parameter estimation. Computers & Security 23(4), 312–319 (2004)

[10] Ingham, K.L., Somayaji, A., Burge, J., Forrest, S., Learning, D.F.A.: representations of HTTP for protecting web applications. Computer Networks 51(5), 1239–1255 (2007)

[11] Halford, W.G.J., Orso, A.: AMNESIA: Analysis and Monitoring for Neutralizing SQL-Injection Attacks. In: Proceedings of the 20th IEEE/ACM International Conference on Automated Software Engineering ASE 2005, Long Beach, CA, pp. 174–183 (2005)
[12] Wassermann, G., Su, Z.: Sound and Precise Analysis of Web Applications for Injection Vulnerabilities. In: Proceedings of the 2007 ACM SIGPLAN Conference on Programming Language Design and Implementation PLDI 2007, San Diego, CA, pp. 32–41 (2007)
[13] Kirda, E., Kruegel, C., Vigna, G., Jovanovic, N.: Noxes: A Client-Side Solution for Mitigating Cross-Site Scripting Attacks. In: Biham, E., Youssef, A.M. (eds.) SAC 2006. LNCS, vol. 4356, pp. 330–337. Springer, Heidelberg (2007)
[14] Kals, S., Kirda, E., Kruegel, C., Jovanovic, N., SecuBat, A.: Web Vulnerability Scanner. In: Proceedings of the 15th International Conference on World Wide Web 2006, Edinburgh, Scotland, pp. 247–256. ACM Press, New York (2006)
[15] Keim, D.A., Mansmann, F., Schneidewind, J., Schreck, T.: Monitoring Network traffic with Radial Analyzer. In: 2006 Symposium On Visual Analytics, Baltimore, MD, pp. 123–128 (2006)
[16] Teoh, S.-T., Ranjan, S., Nucci, A., Chuan, C.-N.: BGP Eye: A New Visualization Tool for Real-time Detection and Analysis of BGP Anomalies. In: Proceedings of the 3rd International Workshop on Visualization for Computer Security VizSEC 2006, Alexandria, Virginia, pp. 81–90 (2006)
[17] Teoh, S.-T., Ma, K.-L., Wu, S.-F., Jankun-Kelly, T.J.: Detecting Flaws and Intruders with Visual Data Analysis. Computer Graphics and Applications 24(5), 27–35 (2004)
[18] Axelsson, S.: Combining a Bayesian Classifier with Visualisation: Understanding the IDS. In: Proceedings of the 2004 ACM workshop on Visualization and data mining for computer security, pp. 99–108. ACM Press, Washington (2004)
[19] Chirillo, J.: The Top 75 Hack Attacks. In: Long, C.A. (ed.) Hack attacks revelead, 2nd edn. Wiley, Indianapolis (2002)
[20] Fingerprinting Port 80 Attacks, A look into web server and web application attack signatures, admin@cgisecurity.com (2002)
[21] Carpenter, G., Grossberg, S.: A Massively Parallel Architecture for a Self-Organizing Neural Pattern Recognition Machine. Computer Vision, Graphics and Image Processing 37, 54–115 (1987)
[22] Xydas, I., Miaoulis, G., Bonnefoi, P.-F., Plemenos, D., Ghazanfarpour, D.: 3D Graph Visualisation of Web Normal and Malicious Traffic. In: Proceedings of the 10th International Conference on Information Visualization (IV 2006), London, UK, pp. 621–629 (2006), doi:10.1109/iv.2006.2.
[23] Haykin, S.: Neural networks, a comprehensive foundation, 2nd edn. Prentice-Hall, Englewood Cliffs (1999)
[24] Montana, D., Davis, L.: Training feedforward neural networks using genetic algorithms. In: Proceedings of 11th International Joint Conference Artificial Intelligence, pp. 762–767. Morgan Kaufmann, San Francisco (1989)
[25] GraphViz software, `http://www.graphviz.org`
[26] Tulip software, `http://www.tulip-software.org`
[27] Xydas, I.: Network security policy surveillance aid using intelligent visual representation and knowledge extraction from a network operation graph, Doctoral dissertation, University of Limoges, France (2007)
[28] Webb, A.: Statistical Pattern Recognition, 2nd edn. Wiley, England (2005)
[29] Hogg, R., Tanis, E.: Probability and Statistical Inference, 7th edn. Pearson Prentice Hall, NJ (2006)

Author Index